Conformal Maps of Nonsmooth Surfaces and Their Applications

I0485366

Vladimir M. Miklyukov

Exlibris Corporation, Philadelphia
2008

Library of Congress Control Number: 2008903546
ISBN: Hardcover   978-1-4363-3693-2
                 Softcover   978-1-4363-3692-5

This book was printed in the United States of America.

To order additional copies of this book, contact:
Xlibris Corporation
1-888-795-4274
www.Xlibris.com
Orders@Xlibris.com
43489

Preface

Astonishingly, under the large number of books on conformal maps between plane domains, the corresponding theory for surfaces is not represented with a connected account up to now.

The aim of this book is the introduction to the theory of conformal maps between non-regular surfaces. We are restricted to examination of locally Lipschitz surfaces. Apparently, such class is minimally necessary in generalizations and reasonably enough for applications.

Below following problems are touched upon: the existence and uniqueness of maps, the boundary behavior of maps and prime ends of surfaces, theorems of the Ahlfors and Warschawski type, applications in a problem of the gas dynamics equation and qualitative questions of the theory of minimal surfaces.

Remark that the considered class of problems does not require applications of complex variables. Thus we can courageously offer another title of this book: "Conformal map without complex variables". Nevertheless, in such places where the application of complex variables is justified by concepts of convenient, we use the language of complex variables. The author attempted to make this presentation maximally simple such that it will be accessible for everybody including young mathematicians started professional work in this domain.

In the book base, there are lectures written by the author for masters of the department of the mathematical analysis and the function theory of Volgograd State University in the 2004/05 academic year. Therefore, as a rule, in the case of the necessary choice between the account of the result in the maximal generality and the showing of a method of its receipt, we preferred the last and refered to find the generality at original articles.[1]

We formulate a series of unsolved problems for beginners.

The author hopes that the book will be useful to readers, interested in conformal mappings and their applications.

Vladimir Michaelovich Miklyukov
Laboratory "Superslow processes",
Volgograd State University,
University avenue 100, Volgograd 400062, RUSSIA
E-mail: miklyuk@mail.ru

---

[1]Rephrasing the well-known aphorism: 'We do not sell fish, we sell fishing-tackles'.

Contents

Chapter 1

Instrumentarium

The chapter provides instrumentarium for investigation of conformal mappings of non-regular surfaces. We introduce a module and capacity of condensers on an abstract surface, establish their estimates and prove that they are coinciding. The Principle of Length and Area on abstract surfaces is proved.

1.1 Abstract Surfaces

We start to terminology. By [52, **1.1**] we shall use the following concept.

A *dilatation* of a map $f : X \to Y$ of a metric space X onto a metric space Y is the quantity

$$\mathrm{dil}\,(f) = \sup_{\substack{x, x' \in X \\ x \neq x'}} \frac{r(f(x), f(x'))}{r(x, x')}.$$

Here r are distances dist_X on X and dist_Y on Y.

A map f is called *Lipschitz* if $\mathrm{dil}\,(f) < \infty$.

A homeomorphism $f : X \to Y$ is called *bi-Lipschitz* if f and f^{-1} are Lipschitz.

If every point $x \in X$ has a neighborhood U such that the restriction $f^* = f|_U$ satisfies the assumption $\mathrm{dil}\,(f^*) < \infty$, then f is called *locally Lipschitz* and *locally bi-Lipschitz* if

$$\mathrm{dil}\,(f^*) < \infty \quad \text{and} \quad \mathrm{dil}\,(f^*)^{-1} < \infty.$$

Let $\mathbb{R}^2$ be a two-dimensional plane, $x = (x_1, x_2)$ be a point of $\mathbb{R}^2$, $S(x, r)$ be a circle of radius $r > 0$ with center $x \in \mathbb{R}^2$, and $B(x, r)$ be a disk with its boundary $\partial B(x, r) = S(x, r)$.

Let $D \subset \mathbb{R}^2$ be a domain. We define an *abstract surface* Ω over a domain D. The surface Ω is given if there are given elements of arc lengths on Ω and its area element.

We denote by $\Gamma(D)$ a set of the Jordan locally rectifiable (with respect to the Euclidean metric) arcs or curves γ lying on D. We shall assume that along every γ, there is given a direction (in particular, from one endpoint to another). Every closed curve $\gamma \in \Gamma(D)$ can be defined in the form

$$x = x(s) = (x_1(s), x_2(s)) : [0, \text{length } \gamma] \to D.$$

Here $0 \le s \le \text{length } \gamma$ is Euclidean arc length, counting from a starting point $x(0)$ up to a moving point $x(s)$ along direction on γ. The locally rectifiable arcs γ can be parameterized by arc lengths counting from a fixed point to a positive and negative directions along γ.

Suppose that along every $\gamma \in \Gamma(D)$, there is given a nonnegative Lebesgue measurable function $h_\gamma(x) : \gamma \to \mathbb{R}$. We shall denote the set of such functions for the arcs $\gamma \in \Gamma(D)$ by $\mathcal{H} = \{h_\gamma\}$.

We shall say that a function collection $\mathcal{H}$ is *coordinated* at $a \in D$ if the values $h_\gamma(a)$ are coinciding for all $\gamma \in \Gamma(D)$ passing through a in the direction of $\xi \in S(a, 1)^1$.

We assume that the collection $\mathcal{H}$ is coordinated for a.e. points on D. Thus there exists an nonnegative function $H(x, \xi)$ defined for a.e. $x \in D$ and all directions $\xi \in S(x, 1)$. We extend H with respect to ξ onto $\mathbb{R}^2$ by the rules $H(x, \lambda \xi) = \lambda H(x, \xi)$, $\lambda = \text{const} \ge 0$. Then a.e. along every $\gamma \in \Gamma(D)$, we shall have

$$H(x, dx) = h_\gamma(x) |dx|. \tag{1.1}$$

Fix a negative function σ defined a.e. in D and Lebesgue measurable.

Below an *abstract surface* Ω is an arbitrary triple (D, H, σ) of described form.

The quantity

$$ds_\gamma = h_\gamma(x) |dx| \tag{1.2}$$

is called an *length element* of $\gamma \in \Gamma(D)$ at $x \in D$, and

$$d\Omega = \sigma(x) \, dx_1 dx_2 \tag{1.3}$$

is called an *area element* of Ω.

For $x', x'' \in D$, we define a *distance*

$$r_\Omega(x', x'') = \inf \int_\gamma ds_\gamma = \inf \int_\gamma H(x, dx) \tag{1.4}$$

_____

[1] that is, having a common tangent vector at a

9

where the infimum is taken over all oriented arcs $\gamma \in \Gamma(D)$ leading from x' to x''.

Since the densities $h_\gamma \in \mathcal{H}$ depend on directions along arcs, generally speaking, $r_\Omega(x', x'') \neq r_\Omega(x'', x')$. Thus an abstract surface simulates an anisotropic medium.

Moreover, an abstract surface can have residual sets of singular points. This property makes possible to simulate mediums with dislocations.[2] For pictures of surfaces with singularities arising in Synergetics, see Haken [57].

For examples of abstract metrics, see in monograph of Misner, Thorne, Wheeler [127] on gravitation and the monograph of Chandrasekhar [27] on mathematical theory of black holes. Interesting samples are contained in books of Panteleev [143] and Asanov [7].

Rich in content examples of abstract surfaces are supplied by the λ-normalized planes of Il'yutko [62]. Unlike $\mathbb{R}^2$, unit circles are non smooth on the λ-normalized plane.

Consider other examples.

Example 1.5 Let $D \subset \mathbb{R}^2$ be a domain and let $E \subset D$ be a closed set of zero linear Hausdorff measure[3]. Let $H(x, \xi) \geq 0$ be a function continuous on $D \setminus E$ for every $\xi \in \mathbb{R}^2$. For an arbitrary arc $\gamma \in \Gamma(D)$, the set $\gamma \cap E$ has zero Hausdorff linear measure, and hence, we can extend $h_\gamma(x) = H|_\gamma$. We choose $\sigma(x) \equiv 1$.

Consider an abstract surface $\Omega = (D, H, 1)$. Below such surfaces are denoted by $(D, H(x, ds))$. Here the distance is the same as the infimum of

$$r_\Omega(x', x'') = \int_\gamma H(x,\, dx)$$

over all arcs γ joining x', x'' in D, and the area of a measurable subset $E \subset D$ is

$$\mathrm{mes}_2\,(E) = \iint_E dx_1 dx_2\,.$$

□

Example 1.6 Let Ω be a surface defined by a locally Lipschitz immersion[4]

$$f : D \subset \mathbb{R}^2 \to \mathbb{R}^m, \quad m \geq 3\,.$$

---

[2] On dislocations, for example, see Godunov, Romenskii [45, Chapter II]).
[3] Here and below, by $\mathrm{mes}_\alpha(E)$ $(0 < \alpha < \infty)$, we denote Hausdorff α-measure of a set E (see, for example, [152, Chapter 2]).
[4] See Section 2.1 below

For every arc $\gamma \in \Gamma(D)$, the contraction $f^* = f|_\gamma$ is locally Lipschitz and hence is absolutely continuous along γ. Thus an length element (1.2) is defined with

$$h_\gamma(x) = \left(\sum_{i,j=1}^{2} g_{ij}(x) \frac{dx_i}{ds} \frac{dx_j}{ds} \right)^{1/2},$$

and

$$g_{11} = \left| \frac{\partial f}{\partial x_1} \right|^2, \quad g_{12} = g_{21} = \left\langle \frac{\partial f}{\partial x_1}, \frac{\partial f}{\partial x_2} \right\rangle, \quad g_{22} = \left| \frac{\partial f}{\partial x_2} \right|^2.$$

In the capacity of σ, we may choose

$$\sigma = \sqrt{g}, \quad g = g_{11}g_{22} - g_{12}^2.$$

$\square$

1.2　Pseudometric

A function $r_\Omega : D \times D \to \mathbb{R}$ is a *pseudometric*. Recall necessary concepts [82, §21].

Let $\mathcal{X}$ be a nonempty set and let $r : \mathcal{X} \times \mathcal{X} \to \mathbb{R}$ be a function with the following properties:

$$(\alpha) \quad r(x,x) \ = 0 \ \text{ and } r(x,y) \geq 0 \ \text{ for all } x, y \in \mathcal{X};$$

$$(\beta) \quad r(x,y) \ \leq r(x,z) + r(z,y) \text{ for all } \ x, y, z \in \mathcal{X}.$$

A pair $(\mathcal{X}, r)$ is a *pseudometric space,* and a function r is a *pseudometric*. Note that here we do not assume the symmetry property of r that is, in general case $r(x,y) \neq r(y,x)$.

On $\mathcal{X}$, we may introduce a topology associated to a pseudometric r that is a topology defined by the system of neighborhoods

$$U_\varepsilon(x) = \{y \in \mathcal{X} : r(x,y) < \varepsilon\}.$$

Thus by standard way, we can define a limit of $f : \mathcal{X} \to \mathbb{R}$ at a point, its continuity, uniform continuity, etc. For example, a sequence $\{x_k\}$, $k = 1, 2, \ldots,$ is fundamental with respect to a pseudometric r if and only if for every $\varepsilon > 0$ there exists $N(\varepsilon)$ such that for any $n > m > N(\varepsilon)$, we have $r(x_n, x_m) < \varepsilon$.

1.3   The Distance r_Ω as a Finsler Metric

Let $D \subset \mathbb{R}^2$ be a domain. We consider an abstract surface $\Omega = (D, H, \sigma)$. Suppose that $H(x, \xi)$ is a function defined for a.e. $x \in D$, all $\xi \in \mathbb{R}^2$ and satisfying the following conditions:

(a) $c_1|\xi| \le H(x, \xi) \le c_2|\xi|$, $c_1(D'), c_2(D') = \text{const} > 0$, a.e. on every subdomain $D' \subset\subset D$;

(b) for a.e. $x \in D$, the set

$$\Xi(x) = \{\xi \in \mathbb{R}^2 : H(x, \xi) < 1\}$$

is convex;

(c) $H(x, \lambda\xi) = \lambda H(x, \xi)$ for $\lambda \ge 0$.

We introduce a dual function

$$G(x, \eta) = \sup_{\xi \in \Xi(x)} \langle \xi, \eta \rangle. \tag{1.7}$$

Here $\langle \xi, \eta \rangle$ is a standard scalar product of vectors ξ and η in $\mathbb{R}^2$.
We set

$$G^+(x) = \sup_{|\eta|=1} \sup_{H(x,\xi)=1} \langle \xi, \eta \rangle.$$

It is not difficult to check that $G(x, \eta)$ satisfies assumptions (a) and (b). Moreover, a.e. on D the following condition holds

$$G(x, \xi) = \sup_{\eta: H(x,\eta) \neq 0} \frac{\langle \xi, \eta \rangle}{H(x, \eta)} \tag{1.8}$$

(see [150, §15]).

In general case, $G(x, \eta)$ has on $D \times \mathbb{R}^2$ values of $\overline{\mathbb{R}}$. The infinity values of $G(x, \eta)$ arise in cases where the convex set $\Xi(x)$ is unbounded. On the other hand, it is not difficult to verify that $\Xi(x)$ is bounded if and only if $G^+(x) < +\infty$.

Example 1.9   Let $\{e_1, e_2\}$ be a standard orthonormal basis in $\mathbb{R}^2$, and $H(x, \xi) = |\langle e_1, \xi \rangle|$. Then

$$\Xi(x) = \{\xi : |\langle e_1, \xi \rangle| < 1\} = \{\xi \in \mathbb{R}^2 : |\xi_1| < 1\}.$$

Here the dual function has the form

$$G(x, \eta) = \begin{cases} |\eta_1| & \text{for } \eta_2 = 0, \\ +\infty & \text{for } \eta_2 \neq 0 \end{cases}$$

and admits infinity values. The function $G^+(x) \equiv +\infty$.

$\square$

Theorem 1.10   *If H satisfies a), b), and c), then r_Ω has $\alpha)$ and $\beta)$ of the pseudometric.*

Proof. The property $\alpha)$ is obvious. We show that the property $\beta)$ holds. Let $x, y, z \in D$, and let

$$r_\Omega(x, z), \quad r_\Omega(z, y) < \infty.$$

We fix $\epsilon > 0$ and choose arcs $\gamma_i(t) : [0, 1] \to \mathbb{R}$, $i = 1, 2$, with

$$\gamma_1(0) = x, \quad \gamma_1(1) = z, \quad \int_0^1 H(\gamma_1(t), \dot{\gamma}_1(t)) \, dt < r_\Omega(x, z) + \frac{\epsilon}{2},$$

and

$$\gamma_2(0) = z, \quad \gamma_2(1) = y, \quad \int_0^1 H(\gamma_2(t), \dot{\gamma}_2(t)) \, dt < r_\Omega(z, y) + \frac{\epsilon}{2}.$$

We let

$$\gamma_3(t) = \begin{cases} \gamma_1(2t) & \text{for } t \in [0, \frac{1}{2}), \\ \gamma_2(2t - 1) & \text{for } t \in [\frac{1}{2}, 1]. \end{cases}$$

Then by homogeneity of $H(x, \xi)$ with respect to ξ,

$$r_\Omega(x, y) \leq \int_0^1 H(\gamma_3(t), \dot{\gamma}_3(t)) \, dt =$$

$$= \int_0^1 H(\gamma_1(t), \dot{\gamma}_1(t)) \, dt + \int_0^1 H(\gamma_2(t), \dot{\gamma}_2(t)) \, dt \leq$$

$$\leq r_\Omega(x, z) + r_\Omega(z, y) + \epsilon.$$

By arbitrariness of $\epsilon > 0$, we arise to the triangle axiom. If $r_\Omega(x, z)$ or $r_\Omega(z, y)$ equals to $+\infty$, then $\alpha)$ is obvious. $\square$

In a *Finsler pseudometric*, an area element is not uniquely defined (see, for example, [151, Chapter I, §8], [7]). Thus, in the general case, choice of the function σ is also ambiguous for Ω.

In the case if $H(x, \xi) = |\xi|$, $\sigma \equiv 1$ and

$$r_\Omega(x, y) = \inf_\gamma \int_0^1 |\gamma'(t)|\, dt = \inf_\gamma \int_\gamma |dx|\,, \qquad (1.11)$$

a metric $r_\Omega(x, y)$ is called *inner* (or *Mazurkiewicz distance*) in D.

On solutions of the equation $H(x, \nabla u) \equiv 1$, see [95] and references.

1.4   The Boundary of Abstract Surfaces

It is useful to describe a boundary of an abstract surface. Generally speaking, a domain boundary is not uniquely defined in topology, and a way of its introduction depends on the problem which is serviced by it. Here we shall consider a simplest way: the boundary as a supplement of Ω with respect to a Finsler pseudometric.

Let D be a domain in $\mathbb{R}^2$. We consider an abstract surface $\Omega = (D, H, \sigma)$. Let $r_\Omega(x', x'')$ be a Finsler pseudometric defined by (1.4).

We denote by D_r the supplement of D with respect to r_Ω. Namely, by D_r, we denote the set of all equivalent classes of the sequences $\{a_n\}$ in D fundamental with respect to r_Ω. Next let $\partial D_r = D_r \setminus D$ be the boundary of D with respect to r.

We interpret the boundary values $\varphi : \partial D_r \to \mathbb{R}$ of f defined on the domain $D \subset \mathbb{R}^2$ as limits of f with respect to r_Ω. Namely for an arbitrary point $x \in \partial D_r$, we set[5]

$$\varphi(x) = \lim_{y \to x,\, y \in D} f(y)\,.$$

In the general case, connections between limiting values of $\varphi : D \to \mathbb{R}$ in the Euclidean sense $\varphi|_{\partial D}$ and in the pseudometric sense $\varphi|_{\partial D_r}$ do not exist.

On comparison to the Euclidean boundary ∂D of $D \subset \mathbb{R}^2$ with ∂D_r, see [50, Section 4].

1.5   Module of Arc Families

We describe a very important concept of a *module of arc family* on an abstract surface. Let $\Omega = (D, H, \sigma)$ be an abstract surface given over a domain $D \subset \mathbb{R}^2$.

---

[5]Here $y \to x$ if and only if $r_\Omega(x, y) \to 0$.

We shall say that a nonnegative locally bounded, measurable by Lebesgue function $\rho \geq 0$ is *admissible* for a subfamily Γ of $\gamma \in \Gamma(D)$ if ρ is measurable along every arc $\gamma \in \Gamma$, and

$$\int_{\gamma} \rho(x)\, ds_{\gamma} \geq 1 \quad \text{for all} \quad \gamma \in \Gamma. \tag{1.12}$$

The quantity

$$\operatorname{mod}_{\Omega}(\Gamma) = \inf_{\rho} \iint_{D} \rho^2\, d\Omega \tag{1.13}$$

is called the *module of the family* Γ. Here the infimum is taken over all functions ρ which are admissible for Γ.

Sometimes instead of module, it is convenient to use a quantity

$$\frac{1}{\operatorname{mod}_{\Omega}(\Gamma)}$$

that is called an *extremal length* of a family Γ.

In the case where the metric $ds_{\gamma} = |dx|$ is Euclidean and the area element $d\Omega = dx_1 dx_2$ (see Example 1.5), we use the simplified designation $\operatorname{mod} \Gamma$.

Theorem 1.14   *For an arbitrary arc family $\Gamma \subset \Gamma(D)$, the following equality*

$$\operatorname{mod}_{\Omega} \Gamma = \inf_{\rho} \frac{\displaystyle\iint_{D} \rho^2(x)\, d\Omega}{\left(\displaystyle\inf_{\gamma \in \Gamma} \int_{\gamma} \rho(x)\, ds_{\gamma} \right)^2} \tag{1.15}$$

holds. Here the infimum is taken over all nonnegative, Lebesgue measurable on D functions ρ.

Proof. Indeed, for every nonnegative, Lebesgue measurable on D function ρ, also measurable along every $\gamma \in \Gamma$, it is easy to see that the function

$$\tilde{\rho} = \frac{\rho(x)}{\displaystyle\inf_{\gamma \in \Gamma} \int_{\gamma} \rho(x)\, ds_{\gamma}}$$

is admissible for Γ. Thus

$$\operatorname{mod}_{\Omega} \Gamma \leq \left(\iint_{D} \rho^2(x)\, d\Omega \right) \left(\inf_{\gamma \in \Gamma} \int_{\gamma} \rho(x)\, ds_{\gamma} \right)^{-2}.$$

Passing to the infimum at the right side of this inequality, by arbitrariness of ρ, we have

$$\text{mod}_\Omega \Gamma \leq \inf_\rho \frac{\displaystyle\iint_D \rho^2(x)\, d\Omega}{\displaystyle\inf_{\gamma \in \Gamma} \int_\gamma \rho(x)\, ds_\gamma}. \tag{1.16}$$

Conversely, let ρ be an admissible function for Γ. Then

$$\inf_{\gamma \in \Gamma} \int_\gamma \rho(x)\, ds_\gamma \geq 1,$$

and hence,

$$\frac{\displaystyle\iint_D \rho^2(x)\, d\Omega}{\left(\displaystyle\inf_{\gamma \in \Gamma} \int_\gamma \rho(x)\, ds_\gamma\right)^2} \leq \iint_D \rho^2(x)\, d\Omega.$$

Passing to the infimum over all admissible functions ρ at the first and next at the right side of this inequality, we obtain

$$\inf_\rho \frac{\displaystyle\iint_D \rho^2(x)\, d\Omega}{\left(\displaystyle\inf_{\gamma \in \Gamma} \int_\gamma \rho(x)\, ds_\gamma\right)^2} \leq \text{mod}_\Omega \Gamma.$$

Comparing with (1.16), we verify that (1.15) holds. $\qquad\qquad\square$

Below we formulate some properties of the introduced quantity known in the Euclidean case [3], [4]. Let $\Omega = (D, H, \sigma)$ be an abstract surface and let Γ_1 and Γ_2 be subfamilies of $\Gamma(D)$. If $\gamma_1 \in \Gamma_1$, $\gamma_2 \in \Gamma_2$, then following by Ahlfors [3, Chapter I, Section **D**], we denote by $\gamma_1 + \gamma_2$ the curve[6], obtained by extension of γ_1 with γ_2. The symbol $\Gamma_1 + \Gamma_2$ means that every curve $\gamma_1 \in \Gamma_1$ extends by a curve $\gamma_2 \in \Gamma_2$ and conversely. The symbol $\Gamma_1 \cup \Gamma_2$ means the standard set-theoretic union of the families. If Γ_1 and Γ_2 are lying on non-overlapping measurable sets, then it is easy to see that

$$\frac{1}{\text{mod}_\Omega\,(\Gamma_1 + \Gamma_2)} \geq \frac{1}{\text{mod}_\Omega\,\Gamma_1} + \frac{1}{\text{mod}_\Omega\,\Gamma_2}, \tag{1.17}$$

---

[6] Here, just as in [3], we extend the curve definition allowing non-connected curves or arcs.

and
$$\mathrm{mod}_\Omega \, (\Gamma_1 \cup \Gamma_2) = \mathrm{mod}_\Omega \, \Gamma_1 + \mathrm{mod}_\Omega \, \Gamma_2 \,. \tag{1.18}$$

For an arbitrary pair Γ_1 and Γ_2, we have

$$\mathrm{mod}_\Omega \, (\Gamma_1 \cup \Gamma_2) \leq \mathrm{mod}_\Omega \, \Gamma_1 + \mathrm{mod}_\Omega \, \Gamma_2 \,. \tag{1.19}$$

The following property is called a *symmetry principle*. Let $\mathbf{H}$ be the upper half-plane on $\mathbb{R}^2$. If Γ is a family of arcs (or curves) on $\mathbf{H}$, then by $\overline{\Gamma}$ we denote the family $\{\overline{\gamma}\}$ of all locally rectifiable arcs (or curves) lying in the lower half-plane and symmetric to $\gamma \in \Gamma$ with respect to the horizontal axis.

Theorem 1.20   *For every arc family $\Gamma \subset \mathbf{H}$, the following property*

$$\mathrm{mod} \, (\Gamma) = 2 \, \mathrm{mod} \, (\Gamma + \overline{\Gamma}) \tag{1.21}$$

holds.

Proof. This statement is a reformulation of Theorem 5 [3, Chapter I]. $\square$

Exercise 1.22 Formulate and prove an analog of Theorem 1.20 for curves on an abstract surface Ω.

1.6   Calculation to the Module

Consider examples of module calculations on the Euclidean plane.

Example 1.23   Let R be a rectangle with lengths of sides a, b which are parallel to coordinate axes Ox_1, Ox_2 respectively. Let Γ be the family of the arcs joining vertical sides (of the length b).

We calculate the module of this family. Let $\rho(x_1, x_2)$ be a function admissible for Γ. Denote by l_y the linearity segment in R, lying in parallel way to the axis Ox_1 at a height of $0 < y < b$. Since ρ is admissible for Γ, then

$$1 \leq \int\limits_{l_y} \rho(x_1, x_2) \, |dx| = \int\limits_0^a \rho(x_1, y) \, dx_1 \,.$$

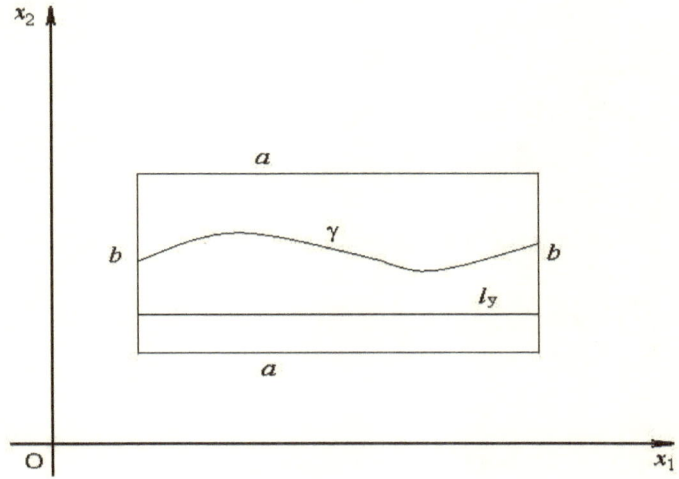

Fig. 1.1.

By the Cauchy inequality,

$$\int\limits_{0}^{a} \rho\, dx_1 \leq \left(\int\limits_{0}^{a} dx_1 \right)^{1/2} \left(\int\limits_{0}^{a} \rho^2(x_1, y)\, dx_1 \right)^{1/2}.$$

Thus

$$\int\limits_{0}^{a} \rho^2(x_1, y)\, dx_1 \geq \frac{1}{a},$$

and we obtain

$$\iint\limits_{R} \rho^2(x_1, x_2)\, dx_1 dx_2 \quad = \int\limits_{0}^{b} dx_2 \int\limits_{0}^{a} \rho^2\, dx_1 \geq$$

$$\geq \frac{1}{a} \int\limits_{0}^{b} dx_2 = \frac{b}{a}.$$

Passing to the infimum over all admissible for Γ functions ρ, we arise to the relation

$$\mod \Gamma \geq \frac{b}{a}\,. \tag{1.24}$$

Now we consider the function

$$\rho_0(x_1, x_2) = \frac{1}{a}\,.$$

Thus function is admissible for Γ, because for every arc $\gamma \in \Gamma$, the inequality

$$\int_\gamma \frac{1}{a}\sqrt{dx_1^2 + dx_2^2} = \frac{1}{a}\int_\gamma \sqrt{dx_1^2 + dx_2^2} = \frac{1}{a}\,\text{length}(\gamma) \geq 1$$

holds.

From this

$$\mod \Gamma \quad \leq \quad \iint_R \rho_0^2 \, dx_1 dx_2 =$$

$$= \quad \iint_R \frac{1}{a^2} dx_1 dx_2 = \frac{b}{a}\,.$$

Comparing it with (1.24), we obtain

$$\mod \Gamma = \frac{b}{a}\,. \tag{1.25}$$

Analogously, for the module of the family $\widetilde{\Gamma}$ of arcs, joining horizontal sides in R, we have

$$\mod \widetilde{\Gamma} = \frac{a}{b}\,,$$

and by (1.25) we arise to the important relation

$$\mod \Gamma \, \mod \widetilde{\Gamma} = 1\,. \tag{1.26}$$

$\square$

Example 1.27   We consider a circular ring

$$K = \{(x_1, x_2) : 0 < r < \sqrt{x_1^2 + x_2^2} < R\}\,.$$

Let Γ be a family of the rectifiable arcs γ lying on K and joining the boundary circles.

We compute the module of this family. Denote by l_θ a segment of a radius, lying between boundary circles and outgoing of the origin with the angle θ to the axis Ox_1. For every function $\rho(x_1, x_2)$ admissible for Γ, we have

$$1 \leq \int_{l_\theta} \rho\, |dx| = \int_r^R \rho(t\cos\theta, t\sin\theta)\, dt \leq$$

$$\leq \left(\int_r^R \frac{dt}{t} \right)^{1/2} \left(\int_r^R \rho^2\, t\, dt \right)^{1/2},$$

and hence,

$$\int_r^R \rho^2\, t\, dt \geq \frac{1}{\ln \dfrac{R}{r}}.$$

Integrating with respect to θ from 0 to 2π, we obtain

$$\iint_K \rho^2(x_1, x_2)\, dx_1 dx_2 = \int_0^{2\pi} d\theta \int_r^R \rho^2(te^{i\theta})\, t\, dt \geq \frac{2\pi}{\ln \frac{R}{r}},$$

and by arbitrariness ρ

$$\operatorname{mod}\Gamma \geq 2\pi \left(\ln \frac{R}{r} \right)^{-1}. \tag{1.28}$$

To prove the inverse inequality, we choose

$$\rho_0(x_1, x_2) = \frac{1}{\sqrt{x_1^2 + x_2^2}} \left(\ln \frac{R}{r} \right)^{-1}.$$

This function is admissible for Γ, because

$$\int_\gamma \frac{1}{\ln \dfrac{R}{r}} \frac{\sqrt{dx_1^2 + dx_2^2}}{\sqrt{x_1^2 + x_2^2}} = \frac{1}{\ln \dfrac{R}{r}} \int_\gamma \frac{|dx|}{|x|} \geq$$

$$\geq \frac{1}{\ln \dfrac{R}{r}} \int_r^R \frac{dt}{t} = 1.$$

Therefore,

$$\operatorname{mod}\Gamma \;\leq\; \iint\limits_{K} \rho_0^2\, dx_1 dx_2 =$$

$$= \frac{1}{\ln^2 \dfrac{R}{r}} \iint\limits_{K} \frac{dx_1 dx_2}{|x|^2} = \frac{2\pi}{\ln \dfrac{R}{r}}\,.$$

Comparing it with (1.28), we arrive to the following relation

$$\operatorname{mod}\Gamma = \frac{2\pi}{\ln \dfrac{R}{r}}\,. \tag{1.29}$$

$\square$

Example 1.30   Let $\widetilde{\Gamma}$ be a family of closed Jordan curves lying on K and separating the boundary circles. We calculate the module of this family.

Denote by C_t, $r < t < R$, the circle $x_1^2 + x_2^2 = t^2$. For every function ρ admissible for $\widetilde{\Gamma}$, we have

$$1 \;\leq\; \int\limits_{C_t} \rho(x)\,|dx| = \int\limits_{0}^{2\pi} \rho(te^{i\theta})\,t\,d\theta \leq$$

$$\leq \left(\int\limits_{0}^{2\pi} \rho^2\, d\theta\right)^{1/2} \left(\int\limits_{0}^{2\pi} t^2\, d\theta\right)^{1/2}.$$

Thus

$$\int\limits_{0}^{2\pi} \rho^2(te^{i\theta})\, d\theta \geq \frac{1}{2\pi t^2}\,,$$

and we obtain

$$\iint\limits_{K} \rho^2 \, dx_1 dx_2 \;=\; \int\limits_{0}^{2\pi} \int\limits_{r}^{R} \rho^2 \, t \, dt d\theta =$$

$$= \int\limits_{r}^{R} t \, dt \int\limits_{0}^{2\pi} \rho^2 \, d\theta \geq$$

$$\geq \frac{1}{2\pi} \int\limits_{r}^{R} \frac{dt}{t} = \frac{1}{2\pi} \ln \frac{R}{r} \,.$$

By arbitrariness of ρ, we arrive to the relation

$$\operatorname{mod} \widetilde{\Gamma} \geq \frac{1}{2\pi} \ln \frac{R}{r} \,. \tag{1.31}$$

Let

$$\rho_0(x_1, x_2) = \left(2\pi \sqrt{x_1^2 + x_2^2} \right)^{-1} \,.$$

For an arbitrary curve $\gamma \in \Gamma$, we have

$$\int\limits_{\gamma} \rho_0 \, |dx| \;=\; \frac{1}{2\pi} \int\limits_{\gamma} \frac{|dx|}{|x|} =$$

$$= \frac{1}{2\pi} \int\limits_{\gamma} \frac{\sqrt{dt^2 + t^2 d\theta^2}}{t} \geq$$

$$\geq \frac{1}{2\pi} \int\limits_{\gamma} \frac{t|d\theta|}{t} \geq 1 \,.$$

From this

$$\operatorname{mod} \widetilde{\Gamma} \leq \frac{1}{(2\pi)^2} \int\limits_{K} \frac{dx_1 dx_2}{|x|^2} = \frac{1}{2\pi} \ln \frac{R}{r} \,.$$

Comparing this inequality with (1.31), we obtain

$$\operatorname{mod} \widetilde{\Gamma} = \frac{1}{2\pi} \ln \frac{R}{r} \,. \tag{1.32}$$

□

We note the relation

$$\operatorname{mod}\widetilde{\Gamma} = \frac{1}{\operatorname{mod}\Gamma} \tag{1.33}$$

which follows from (1.29) and (1.32).

□

1.7   Module for Finsler Metrics

As in Example 1.27, let K be a circular ring with radii $0 < r < R < \infty$. Let $\Gamma \subset \Gamma(K)$ be a family of locally rectifiable arcs on K joining the boundary circles.

Denote by $l_\theta \in \Gamma$ an radial segment lying between the boundary circles and forming an angle θ with the axis Ox_1.

We assume that over the domain K an abstract surface $\Omega = (D, H, \sigma)$ is defined. We evaluate $\operatorname{mod}_\Omega \Gamma$.

Choose an admissible function $\rho(x)$ for Γ. For an arbitrary arc $l_\theta \in \Gamma$ by (1.1), we have

$$1 \le \int\limits_{l_\theta} \rho \, ds_{l_\theta} = \int\limits_r^R \rho(\tau \, e^{i\theta}) \, H(\tau \, e^{i\theta}, \, e^{i\theta}) \, d\tau$$

where $x = \tau \, e^{i\theta}$.

From this and from the inequality

$$\int\limits_r^R \rho H(\tau \, e^{i\theta}, \, e^{i\theta}) \, d\tau \le \left(\int\limits_r^R \frac{H^2 \, d\tau}{\tau \sigma} \right)^{1/2} \left(\int\limits_r^R \rho^2 \, \sigma(\tau \, e^{i\theta}) \, \tau \, d\tau \right)^{1/2},$$

it follows that

$$\int\limits_K \rho^2 \, \sigma \, dx_1 dx_2 = \int\limits_0^{2\pi} d\theta \int\limits_r^R \rho^2 \sigma \tau \, d\tau \ge \int\limits_0^{2\pi} d\theta \left/ \int\limits_r^R \frac{H^2 \, d\tau}{\tau \, \sigma} \right. .$$

By the arbitrariness of ρ, we conclude that

$$\int\limits_0^{2\pi} d\theta \left/ \int\limits_r^R \frac{H^2(\tau \, e^{i\theta}, \, e^{i\theta}) \, d\tau}{\tau \, \sigma(\tau \, e^{i\theta})} \right. \le \operatorname{mod}_\Omega \Gamma. \tag{1.34}$$

Analogously, for the family of closed Jordan curves $\widetilde{\Gamma}$ separating the boundary circles of the ring K, we have

$$\int\limits_{r}^{R} \frac{d\tau}{\tau} \bigg/ \int\limits_{0}^{2\pi} \frac{H^2(\tau\,e^{i\theta},\,e^{i\theta})\,d\theta}{\sigma(\tau\,e^{i\theta})} \leq \mathrm{mod}_\Omega\widetilde{\Gamma}\,. \tag{1.35}$$

Thus we arrive to the following statement.

Theorem 1.36 *If a domain K is a circular ring of the described form and $\Omega\,(K,H,\sigma)$ is an abstract surface over K, then for the module of the family Γ of arcs joining the boundary circles in K and for the module of the family $\widetilde{\Gamma}$ of closed Jordan curves separating the boundary circles in K, there are the estimates (1.34) and (1.35) respectively.*

Remarks 1.37 Assume that

$$H(x,\xi) = \left(\sum_{i,j=1}^{2} g_{ij}(x)\,\xi_i\xi_j \right)^{1/2},$$

and the coefficients g_{ij} $(i,j=1,2)$ are measurable in K, and for every subdomain $K' \subset\subset K$ for a.e. $x \in K'$ and $\xi \in \mathbb{R}^2$, the following relations hold

$$c_1(K')\,|\xi| \leq H(x,\xi) \leq c_2(K')\,|\xi| \tag{1.38}$$

where $0 < c_1(K'),\, c_2(K') < \infty$ are constants.

Under given assumptions we are able to show estimates for $\mathrm{mod}_\Omega\Gamma$. Below in Theorem 3.49, we will show that by means of a quasi-conformal homeomorphism $y = \varphi(x) : K \to \mathbb{R}^2$, the metric ds_Ω^2 can be transformed to a conformal Euclidean form, i.e. to a metric of the following form

$$ds^2 = \lambda^2(y)\,|dy|^2$$

where $\lambda(y)$ is a positive measurable function.

Under such mapping for every family of arcs $\Gamma \subset K$, we have

$$\mathrm{mod}_\Omega\Gamma = \mathrm{mod}\,\varphi(\Gamma)\,. \tag{1.39}$$

In the case of the conformal Euclidean metric by (1.33) we obtain

$$\mathrm{mod}\,\varphi(\Gamma) = 1/\mathrm{mod}\,\widetilde{\varphi(\Gamma)}\,.$$

From this by (1.39), we conclude that there is an analogous relation for the module in the metric $d_\Omega s^2$, and using by (1.34) and (1.35), we arrive to the estimate for $\text{mod}_\Omega\Gamma$:

$$\left(\int\limits_{r}^{R} \frac{d\tau}{\tau \int\limits_{0}^{2\pi} \dfrac{H^2(\tau\, e^{i\theta},\, e^{i\theta})\, d\theta}{\sigma(\tau\, e^{i\theta})}} \right)^{-1} \geq \text{mod}_\Omega\Gamma \geq \int\limits_{0}^{2\pi} \frac{d\theta}{\int\limits_{r}^{R} \dfrac{H^2(\tau\, e^{i\theta},\, e^{i\theta})\, d\tau}{\tau\, \sigma(\tau\, e^{i\theta})}}.$$

For $H(x,\xi) = |\xi|$ and $\sigma \equiv 1$, the quantities in the left and the right sides of the inequality are equal. Thus we arrive to the well-known formula (1.29) for the module of a circular ring in Euclidean plane.

$$\square$$

Open questions 1.40   Prove the analogous two-sided estimate for $\text{mod}_\Omega\Gamma$ without the assumption on the special form $H(x,\xi)$ and with no employment of auxiliary quasi-conformal mappings.

1.8   Condensers on Surfaces

Let $P, Q \subset D$ be nonempty, closed in a domain $D \subset \mathbb{R}^2$, non-overlapping subsets. A triple $(P, Q; D)$ is called a *condenser*.

We define *the module and the capacity* of the condenser $(P, Q; D)$.

We consider an abstract surface $\Omega = (D, H, \sigma)$. Let $G(x, \eta)$ be the dual to $H(x, \xi)$ function defined by (1.7).

By a module (or Ω-module) of a condenser $(P, Q; D)$ on a surface $\Omega = (D, H, \sigma)$, we shall call the module of the family $\Gamma(P, Q; D) \subset \Gamma(D)$ of the arcs γ lying in D and joining the sets P and Q. Namely, we put

$$\text{mod}_\Omega(P, Q; D) = \text{mod}_\Omega\Gamma(P, Q; D). \tag{1.41}$$

By a capacity (or Ω-capacity) of a condenser $(P, Q; D)$ on a surface $\Omega = (D, H, \sigma)$, we call the following quantity

$$\text{cap}_\Omega(P, Q; D) = \inf \int\limits_{D} G^2(x, \nabla\varphi)\, \sigma\, dx_1 dx_2 \tag{1.42}$$

where the infimum is taken over all functions φ locally Lipschitz on D with $\varphi|_P = 0$, $\varphi|_Q = 1$.

The module and the capacity of the condenser are dual quantities. They are equal in the cases of Euclidean (see Fuglede [40]) and Riemannian (see Miklyukov [111], Dymchenko [32]) metrics and are expressed one through another under the suitable choice of H, G and σ in the case of Finsler metric (see Tkachev, Ushakov [184]). The choice of the module or the capacity in a research is dictated as a rule by reasons of comfort.

Theorem 1.43   *Let $\Omega = (D, H, \sigma)$ be an abstract surface. If $\Delta \subset D$ is a subdomain, then*

$$\operatorname{cap}_\Omega(P,\ Q;\ \Delta) = \operatorname{mod}_\Omega(P,\ Q;\ \Delta). \tag{1.44}$$

Proof. Let $\varphi(x)$ be a function admissible in the variational problem (1.42). We put

$$\Lambda(x) = \varlimsup_{y \to x} \frac{|\varphi(y) - \varphi(x)|}{|y - x|}.$$

By Rademacher Theorem[7] $\varphi(x)$ is differentiable a.e. on D ([37, **3.1.6.**]). We put

$$H_-(x) = \min_{|\xi|=1} H(x, \xi).$$

Denote by $\rho(x)$ a function which coincides with $G(x, \nabla\varphi)$ at the points of differentiability of $\varphi(x)$ belonging to Δ, with the function $\Lambda(x)/H_-(x)$ at points of non-differentiability of $\varphi(x)$ belonging to Δ and vanishing on $D \setminus \Delta$.

The function $\rho(x)$ is admissible for $\Gamma(P, Q; \Delta)$. Indeed, from (1.8) it follows that at every point where $\varphi(x)$ is differentiable, we have

$$|\varphi_{x_1}\, dx_1 + \varphi_{x_2}\, dx_2| \ = |\langle \nabla\varphi, dx\rangle| \le$$

$$\le G(x, \nabla\varphi)\, H(x, dx),$$

and consequently, at every point where $\varphi(x)$ is differentiable:

$$|d\varphi(x)| \le \rho(x)\, H(x, dx). \tag{1.45}$$

---

[7]Theorem (Rademacher).   *Every locally Lipschitz map $f : D \to \mathbb{R}^n$ is differentiable a.e. on the domain D* [37, Theorem **3.1.6**]. *There are more general statements. Theorem (Stepanov). If f satisfies a.e. on a domain D the condition $\varlimsup_{y \to x} |f(y) - f(x)|/|y - x| < \infty$, then f is differentiable a.e. on D* [37, Theorem **3.1.9**].

At every point where $\varphi(x)$ is non-differentiable, we have

$$\Lambda(x)\,|dx| = \rho(x)\,H_-(x)\,|dx| \le \rho(x)\,H(x,\frac{dx}{|dx|})\,|dx|\,.$$

From this and from (1.45) for each arc $\gamma \in \Gamma(P,Q;\Delta)$ it follows

$$\int\limits_{\gamma} \rho\,ds_\Omega \ge \int\limits_{\gamma} |d\varphi(x)| \ge 1\,.$$

Thus

$$\mathrm{mod}_\Omega(P,Q;\Delta) \le \iint\limits_{\Delta} G(x,\nabla\varphi)\,\sigma(x)\,dx_1\,dx_2\,,$$

and we obtain

$$\mathrm{mod}_\Omega(P,Q;\Delta) \le \mathrm{cap}_\Omega(P,Q;\Delta)\,.$$

Now we prove the opposite inequality. We suppose that $\rho(x)$ is admissible for $\Gamma(P,Q;\Delta)$ and satisfied (1.12) for every arc on Δ joining P and Q. We put

$$\varphi^*(x) = \int\limits_{\gamma(x)} \rho\,ds_\Omega$$

where the infimum is taken over all arcs $\gamma(x)$, joining a point $x \in \Delta$ with the set P on Δ.

Because $\rho(x)$ is locally bounded on Δ, then the function

$$\varphi(x) = \min\{1, \varphi^*(x)\}$$

belongs to the class $\mathrm{Lip}D$. The assumption (1.12) implies that $\varphi(x)$ equals 0 and 1 on P and Q respectively. If we show that a.e. on Δ it is fulfilled

$$G(x,\nabla\varphi) \le \rho(x)\,, \tag{1.46}$$

then

$$\mathrm{cap}_\Omega(P,Q;\Delta) \le \iint\limits_{\Delta} \rho^2\,\sigma dx_1 dx_2$$

whence

$$\mathrm{cap}_\Omega(P,Q;\Delta) \le \mathrm{mod}_\Omega(P,Q;\Delta)\,,$$

and (1.44) will be proved.

We prove (1.46). At first we remark that $\varphi(x)$ is differentiable a.e. on D and, by well-known Lebesgue theorem for a.e. $y \in D$, we have

$$\lim_{r\to 0} \frac{1}{r^2} \iint\limits_{|x-y|<r} |\rho(x) - \rho(y)|\,dx_1\,dx_2 = 0\,. \tag{1.47}$$

Thus it is sufficient to verify (1.46) on the set where both noted properties are fulfilled. Let y be a point of this set at which $0 < \varphi(y) < 1$, and let $\gamma(y, \theta)$ be a line segment with the length h outgoing from y with the angle θ to the line $x_2 = 0$. Then

$$|\varphi_{x_1}(y) \cos \theta + \varphi_{x_2}(y) \sin \theta| \leq \lim_{h \to 0} \frac{1}{h} \int_{\gamma(y,\theta)} \rho \, ds_F \,. \tag{1.48}$$

Now we have

$$\frac{1}{h} \int_{\gamma(y,\theta)} \rho \, ds_\Omega \;\leq \rho(y) \frac{1}{h} \int_{\gamma(y,\theta)} H(\tau \, e^{i\theta}, e^{i\theta}) \, d\tau +$$

$$+ \frac{1}{h} \int_{\gamma(y,\theta)} |\rho(\tau \, e^{i\theta}) - \rho(y)| \, H(\tau \, e^{i\theta}, e^{i\theta}) \, d\tau \,. \tag{1.49}$$

Setting h as sufficiently small and integrating by θ, we have

$$\int_0^{2\pi} d\theta \int_{\gamma(y,\theta)} |\rho(\tau \, e^{i\theta}) - \rho(y)| \, H(\tau \, e^{i\theta}, e^{i\theta}) \, d\tau \leq$$

$$\leq c_1 \iint_{|x-y|<h} |\rho(x) - \rho(y)| \frac{dx_1 dx_2}{|x-y|} \leq$$

$$\leq c_2 h^{1/3} \left(\iint_{|x-y|<h} |\rho(x) - \rho(y)|^3 dx_1 dx_2 \right)^{1/3}$$

where c_1, c_2 are constants. Taking into account the local boundedness of $\rho(x)$ and the equality (1.47), we obtain

$$\lim_{h \to 0} \frac{1}{h} \int_0^{2\varphi} d\theta \int_{\gamma(y,\theta)} |\rho(x) - \rho(y)| ds_F = 0 \,,$$

and for a.e. $\theta \in [0, 2\pi]$:

$$\lim_{h \to 0} \frac{1}{h} \int_{\gamma(y,\theta)} |\rho(x) - \rho(y)| \, ds_F = 0 \,.$$

Thus from (1.48) and (1.49) for a.e. directions and, consequently, for all directions θ at y, we have

$$|\varphi_{x_1}\cos\theta + \varphi_{x_2}\sin\theta| \le \rho(y)\,H(y,e^{i\theta})\,.$$

Comparing this relation with (1.8), we see that (1.46) is true for $0 < \varphi(y) < 1$.

Now let R be the set of the points $x \in D$ in which $\varphi(x) = 1$, and let T be its subset such that the contingent of R at every its point is not all plane. The set R has zero measure (see [161, p. 384]), and at every point of differentiability $\varphi(x)$ lying in $R \setminus T$, it is fulfilled $\nabla\varphi = 0$, and consequently, (1.46) is true.

Analogous arguments are suitable for the points of Q where $\varphi(x) = 0$, and (1.46) is proved completely.　　　　　　　　　　　　　　　　　　　　　　　□

1.9　Length and Area Principle

Below we prove a special version of the famous 'Length and Area Principle' which is widely used in the theory of functions and mappings with generalized derivatives (see, for example, Courant [29, Chapter I, §**4**], Lelong-Ferrand [91, p. 7], Suvorov [177, p. 252 – 260], Miklyukov [115], Kufarev [79]). The following version has been proved by Klyachin and Miklyukov.

We consider an abstract surface $\Omega = (D, H, \sigma)$ given over a domain $D \subset \mathbb{R}^2$. Let P, Q be subsets of D closed with respect D and such that $P \cap Q = \emptyset$. Let $\varphi : G \to \mathbb{R}$ be a locally Lipschitz function with properties: $\varphi|_P = 0$, $\varphi|_Q = 1$ and

$$\operatorname*{ess\,inf}_{U} |\nabla\varphi| > 0 \quad \text{for every} \quad U \subset\subset D\,. \tag{1.50}$$

Denote by $E_t = \{x \in D : \varphi(x) = t\}$ the t-level set of the function φ. By [37, Theorem **3.2.15**] the sets E_t are locally rectifiable for a.e. t. We assume that E_t are connected for each t.

Let $G(x, \eta)$ be a function dual to $H(x, \xi)$. We put

$$\zeta^{\pm} = (\pm\varphi_{x_2}, \mp\varphi_{x_1})\,/|\nabla\varphi| \,, \quad G_{0,\varphi}(x) = \max\{G(x,\zeta^{+}), G(x,\zeta^{-}\}\,.$$

The following theorem expresses the "Length and Area Principle" at the metric of the surface Ω.

Theorem 1.51　　*Let $\Omega = (D, H, \sigma)$ be an abstract surface. Then for every locally Lipschitz function $f : D \to \mathbb{R}$, it is fulfilled*

$$\int_{0}^{1} \frac{\operatorname{osc}^2(f; E_t)\,dt}{\displaystyle\int_{E_t} H^2(x, \nabla\varphi)|\nabla\varphi|^{-1}\,\sigma\,|dx|} \le \iint_{D\setminus(P\cup Q)} G^2(x, \nabla f)\sigma(x)\,dx_1 dx_2\,. \tag{1.52}$$

In particular, if

$$\sigma(x) = \frac{G_{0,\varphi}(x)\,|\nabla\varphi(x)|}{H(x,\nabla\varphi)},$$

then

$$\inf_{\varphi}\ \inf_{0\le t\le 1}\ \mathrm{osc}^2(f; E_t) \le \mathrm{cap}_{\Omega}\,(P, Q; D) \iint_D G^2(x, \nabla f)\sigma(x)\,dx_1dx_2. \qquad (1.53)$$

Proof of (1.52) is not difficult. Using (1.8) and observing that a.e. on E_t

$$\frac{dx}{|dx|} = (\pm\varphi_{x_2}, \mp\varphi_{x_1})\,/|\nabla\varphi| = \zeta^{\pm},$$

we have

$$\mathrm{osc}(f; E_t) \ \le \int_{E_t} |\langle \nabla f, dx\rangle| \le$$

$$\le \int_{E_t} G(x, \nabla f)\,H(x, dx) \le$$

$$\le \left(\int_{E_t} H^2(x, \nabla\varphi)\,|\nabla\varphi|\,\frac{|dx|}{\sigma} \right)^{1/2} \left(\int_{E_t} G^2(x, \nabla f)\,\frac{\sigma\,|dx|}{|\nabla\varphi|} \right)^{1/2}.$$

By integration with respect to t, we obtain

$$\int_0^1 \frac{\mathrm{osc}^2(f; E_t)\,dt}{\displaystyle\int_{E_t} H^2(x, \nabla\phi)\,|\nabla\varphi|\,\frac{|dx|}{\sigma}} \le \int_0^1 dt \int_{E_t} G^2(x, \nabla f)\,\frac{\sigma\,|dx|}{|\nabla\varphi|}. \qquad (1.54)$$

Now we use the well-known Kronrod-Federer formula for co-area

$$\int_0^1 dt \iint_{E_t} A(x)\,\frac{|dx|}{|\nabla\varphi|} = \int_{D\setminus(P\cup Q)} A(x)\,dx_1dx_2 \qquad (1.55)$$

which is valid for a Lebesgue measurable function $A : D \setminus (P \cup Q) \to \mathbb{R}$ and every locally Lipschitz function $\varphi : D \to \mathbb{R}$ with (1.50) (for example, see [37, Theorem **3.2.5**]).

By (1.55) we arrive to the following relation

$$\int_0^1 dt \int_{E_t} G^2(x, \nabla f) \frac{\sigma |dx|}{|\nabla \varphi|} = \iint_{D \setminus (P \cup Q)} G^2(x, \nabla f) \, \sigma \, dx_1 dx_2 \,.$$

This relation with (1.54) imply (1.52).

For the proof of (1.53), we use a special choice of σ. We have

$$G_{0,\varphi}^2(x) |\nabla \varphi| \frac{|dx|}{\sigma} = G^2(x, \nabla \varphi) |\nabla \varphi|^{-1} \sigma |dx| \,.$$

Hence by (1.54), we obtain

$$\int_0^1 \frac{\operatorname{osc}^2(f; E_t) \, dt}{\int_{E_t} G^2(x, \nabla \varphi) |\nabla \varphi|^{-1} \sigma |dx|} \leq \int_0^1 dt \int_{E_t} G^2(x, \nabla f) \frac{\sigma |dx|}{|\nabla \varphi|} \,. \tag{1.56}$$

Now we prove that (1.52) implies (1.53). Using the Cauchy inequality and the co-area formula, we have

$$1 \quad \leq \int_0^1 dt \int_{E_t} G^2(x, \nabla \varphi) |\nabla \phi|^{-1} \sigma |dx| \cdot \int_0^1 \frac{dt}{\int_{E_t} G^2(x, \nabla \varphi) |\nabla \varphi|^{-1} \sigma |dx|} =$$

$$= \iint_{D \setminus (P \cup Q)} G^2(x, \nabla \phi) \sigma(x) dx_1 dx_2 \cdot \int_0^1 \frac{dt}{\int_{E_t} G^2(x, \nabla \phi) |\nabla \varphi|^{-1} \sigma} |dx| \,.$$

Thus by (1.52), we obtain

$$\inf_{0 \leq t \leq 1} \operatorname{osc}^2(f; E_t) \quad \leq \int_D G^2(x, \nabla \phi) \sigma(x) dx_1 dx_2 \times$$

$$\times \int_{D \setminus (P \cup Q)} G^2(x, \nabla f) \sigma(x) \, dx_1 dx_2 \,. \tag{1.57}$$

Passing in the right side of (1.57) to the infimum over all functions φ of the described form, we arrive to (1.53). $\qquad \square$

Open questions 1.58 1) Find conditions (necessary, sufficient) of the vanish Ω-capacity of a set, analogous to corresponding results for the (l, p)-capacity [46, **5.3**]. 2) Study the boundary ∂D_{r_Ω} of the domain $D \subset \mathbb{R}^2$. 3) Find conditions of the vanish Ω-capacity of the boundary set $E \subset \partial D_{r_\Omega}$. 4) Research possibilities of an isometric immersion of an abstract surface into a pseudo-Euclidean space. (See Borisenko [22, Section **5**] on the problem of isometric immersions of space forms.)

Chapter 2

Locally Minimal Surfaces

In this chapter we define main concepts concerning to minimal surfaces in Euclidean space. We prove that a height function of a locally Lipschitz and locally minimal surface is an generalized solution of a Laplace-Beltrami equation at the metric of the surface.

2.1 Surfaces in $\mathbb{R}^m$

Let $D \subset \mathbb{R}^2$ be a simply connected domain and let Ω be a two-dimensional surface in $\mathbb{R}^m$, $m \geq 2$, given by a locally Lipschitz vector function

$$\xi = f(x_1, x_2) = (f_1(x), \dots, f_m(x)) : D \to \mathbf{R}^m . \tag{2.1}$$

In the general case, the surface Ω can have self-intersections. We shall say that a surface Ω is *embedded* to $\mathbb{R}^m$ if the vector function f realizes a homeomorphic mapping of the domain D onto the set $f(D)$ with the metric (and, consequently, the topology !) which is induced of $\mathbb{R}^m$. A surface Ω is *immersed* to $\mathbb{R}^m$ if the vector function f has the described property locally in D.

It is clear that if Ω is bi-Lipschitz, then it is embedded, and if Ω is locally bi-Lipschitz, then it is immersed.

Because the vector function f is locally Lipschitz, then by the Rademacher theorem, the differential $df(x)$ exists a.e. on D.

Let $(x_1, x_2) \in D$ be a point in which f is differentiable. By the symbol

$$f' = \left(\begin{array}{cccc} f'_{1x_1} & f'_{2x_1} & \cdots & f'_{mx_1} \\ \\ f'_{1x_2} & f'_{2x_2} & \cdots & f'_{mx_2} \end{array} \right) ,$$

we denote the derivative f at $x = (x_1, x_2)$ if this derivative exists. Using standard designations

$$g_{11} = \left| \frac{\partial f}{\partial x_1} \right|^2, \quad g_{12} = g_{21} = \left\langle \frac{\partial f}{\partial x_1}, \frac{\partial f}{\partial x_2} \right\rangle, \quad g_{22} = \left| \frac{\partial f}{\partial x_2} \right|^2,$$

we define the first quadratic form of Ω on the domain D

$$ds_\Omega^2 = g_{11}\, dx_1^2 + 2g_{12}\, dx_1\, dx_2 + g_{22}\, dx_2^2\,.$$

The first quadratic form ds_Ω^2 has the simplest view in isothermal coordinates. However, the question — What are isothermal coordinates on non-smooth surfaces ? — has no good answer today. We will use the following definition.

Definition 2.2   *Let Ω be a surface given over a domain $D \subset \mathbb{R}^2$ by a locally Lipschitz vector function (2.1). The variables $x = (x_1, x_2)$ are called isothermal coordinates on Ω if*

$$g_{11}(x) = g_{22}(x)\,, \quad g_{12}(x) = 0 \quad a.e.\ on \quad D\,. \tag{2.3}$$

In the case if $x = (x_1, x_2)$ are isothermal coordinates on Ω, we have

$$ds_\Omega^2 = \lambda(x)\,(dx_1^2 + dx_2^2)$$

a.e. on D. Here $\lambda = g_{11} = g_{22}$.

The first condition of (2.3) means that the tensions of f along lines x_1, $x_2 = $ const coincide at points where df exists. The second condition implies mutual orthogonality the images of these lines at corresponding points $\xi = f(x_1, x_2)$. Thus at every point $x \in D$ where f is differentiable and (2.3) are true, the mapping $f : D \to \Omega$ preserve angles between curves, i.e. f is conformal.

The isothermal coordinates existence is researched incompletely for non-regular surfaces. For the $W^{l,p}$-surfaces with $l \geq 3$ and $p > 2$, the isothermal coordinates existence theorem can be obtained from well-known results of B. Bojarskii and I. Vekua [189, §6, Chapter II]). Closed results for Lipschitz graphs belong to Ch.B. Morrey [133], and for locally Lipschitz graphs they can be obtained from results L. Bers [21], P.P. Belinskii [17, Theorem 9], O. Martio and V.M. Miklyukov [98] (under different assumptions with respect domains of graphs).

The general case of nonparametric locally Lipschitz surfaces was considered in [118]. We shall return to this problem below, in Chapter 3.

2.2   The Laplace-Beltrami Equation

Let Ω be a surface immersed in $\mathbb{R}^m$ by a locally Lipschitz vector - function
(2.1). We put

$$g = \det\left(g_{ij}\right).$$

By the Cauchy inequality,

$$g_{12}^2 \leq g_{11}g_{22}.$$

Therefore, we have

$$g = g_{11}g_{22} - g_{12}^2 \geq 0.$$

The area element of Ω is defined as

$$d\sigma_\Omega = \sqrt{g}\,dx_1\,dx_2.$$

If $g(x) > 0$ at a point $x = (x_1, x_2) \in D$, then it is defined the inverse matrix
$(g^{ij}) = (g_{ij})^{-1}$. Thus, if $g(x) > 0$ a.e. on the domain D, then we are right to
say that the Laplace operator with respect to the metric ds_Ω. Namely, we put

$$\Delta_\Omega h \equiv \frac{1}{\sqrt{g}}\sum_{i=1}^{2}\frac{\partial}{\partial x_i}\left(\sqrt{g}\sum_{j=1}^{2}g^{ij}(x)\frac{\partial h}{\partial x_j}\right) = 0. \tag{2.4}$$

The equation (2.4) is called the *Laplace-Beltrami equation* with respect to
the metric ds_Ω^2 (see the motivation M. Schiffer and D.C. Spencer [163, §1,
Chapter 1]).

It is convenient to use the following definition of generalized solutions of the
Laplace equation in the metric ds_Ω^2.

Definition 2.5   *A locally Lipschitz function $h : D \to \mathbb{R}$ is called a generalized
solution of (2.4) (or a harmonic function with respect to a metric ds_Ω) if for
an arbitrary Lipschitz function $\varphi : D \to \mathbb{R}$ with a compact support $\operatorname{supp}\varphi \Subset D$,
it is fulfilled*

$$\iint\limits_{D}\sum_{i,j=1}^{2}g^{ij}\varphi'_{x_i}\,h'_{x_j}\,\sqrt{g}\,dx_1\,dx_2 = 0. \tag{2.6}$$

We define subharmonic and superharmonic functions with respect to a metric ds_Ω.

Definition 2.7 *A locally Lipschitz function $h : D \to \mathbb{R}$ is called a generalized subsolution (supersolution) of (2.4) (or subharmonic, respectively superharmonic function with respect to a metric ds_Ω) if for an arbitrary non-negative (non-positive) Lipschitz function $\varphi : D \to \mathbb{R}$ with a compact support $\operatorname{supp} \varphi \Subset D$, it is fulfilled*

$$\iint\limits_{D} \sum_{i,j=1}^{2} g^{ij} \varphi'_{x_i} h'_{x_j} \sqrt{g}\, dx_1\, dx_2 \leq 0. \qquad (2.8)$$

Let's explain what has been said. Let D be a bounded domain with a smooth boundary. Suppose that $h \in C^2(D)$ is a generalized subsolution of (2.4) with respect to a metric ds_Ω with $C^1(D)$-coefficients. By the Green formula for every function $\varphi \in C^1(\overline{D})$, we have

$$\int\limits_{\partial D} \varphi \sum_{i,j=1}^{2} g^{ij} \mathbf{n}_{x_i} h'_{x_j} \sqrt{g}\, |dx| = \iint\limits_{D} \sum_{i,j=1}^{2} g^{ij} \varphi'_{x_i} h'_{x_j} \sqrt{g}\, dx_1\, dx_2 +$$

$$+ \iint\limits_{D} \varphi \Delta_\Omega h \sqrt{g}\, dx_1\, dx_2$$

where $\mathbf{n} = (\mathbf{n}_{x_1}, \mathbf{n}_{x_2})$ is an unit vector of the exterior normal to the boundary ∂D.

If h is a subsolution and $\varphi|_{\partial D} = 0$, then the assumption (2.8) implies

$$\iint\limits_{D} \varphi \Delta_\Omega h \sqrt{g}\, dx_1\, dx_2 \geq 0.$$

Because this relation holds for each nonnegative function φ, then by the Main Lemma of the variational calculus, we obtain

$$\Delta_\Omega h \geq 0.$$

Therefore, *every subsolution in the classic sense of the Laplace-Beltrami equation is a generalized subsolution.*

Analogously, the cases of generalized supersolutions and generalized solutions can be considered.

2.3   The Height Function

Below we assume that

$$\frac{1}{g} \in L^{\frac{3}{2}}_{\text{loc}}(D). \qquad (2.9)$$

In particular, it means that $g > 0$ a.e. on D.

Let g^{ij} be coefficients of the inverse matrix $(g^{ij}) = (g_{ij})^{-1}$. The supposition (2.9) implies that

$$(g^{ij}) = \frac{\mathrm{adj}(g_{ij})}{g}$$

a.e. on D.

Locally Lipschitz surfaces in $\mathbb{R}^m$ have many interesting (and very exotic !) properties. In particular, their inner and exterior geometries can be not coordinated (see, for example, Nash [135] and Kuiper [78]).

A surface (2.1) is called locally minimal if for every subdomain $D' \subset\subset D$ with a Jordan boundary $\partial D'$ and every Lipschitz vector function

$$\xi = \varphi(x_1, x_2) : D \to \mathbb{R}^m, \quad \varphi|_{D \backslash D'} = f,$$

the following property holds

$$\iint\limits_{D} \sqrt{\left|f'_{x_1}\right|^2 \left|f'_{x_2}\right|^2 - \left\langle f'_{x_1}, f'_{x_2} \right\rangle^2}\, dx_1 dx_2 \leq$$

$$\leq \iint\limits_{D} \sqrt{\left|\varphi_{x_1}\right|^2 \left|\varphi_{x_2}\right|^2 - \left\langle \varphi_{x_1}, \varphi_{x_2} \right\rangle^2}\, dx_1 dx_2 .$$

The following statement is main in this chapter. In particular, it guarantees applicability of standard methods of the partial differential equation theory to height functions. It is the nodal moment in use of direct methods of the variational calculus in the Plateau problem (Lebesgue [89], McShane [104]). To a certain extent, it outlines boundaries of justice of Courant's remark on advantages of parametric methods in the Plateau problem comparatively with direct methods (see [29, Chapter III, §1]), the more so, as the natural class of minimal surfaces in $\mathbb{R}^m$ contains non-smooth surfaces (see, for example, [39, Theorem **24.1.1**], [44, Theorem **11.8**]), or [132, Chapter 8].

Theorem 2.10 *If $\Omega \subset \mathbb{R}^m$ is a two-dimensional locally minimal and locally Lipschitz surface, given by a vector function (2.1) with (2.9), then for every unit vector $\mathbf{e}$, a height function $h(m) = \langle f(m), \mathbf{e} \rangle$ is harmonic in the generalized form with respect to the metric ds^2_Ω.*

Proof. We follow [121]. Because a height function is a linear combination of the components f_i ($i = 1, \ldots, n$) of the vector function $\xi = f(x) : D \to \mathbb{R}^m$, it is sufficient to prove that the component $f_1 = f_1(x)$ is harmonic with respect to the metric ds^2_Ω.

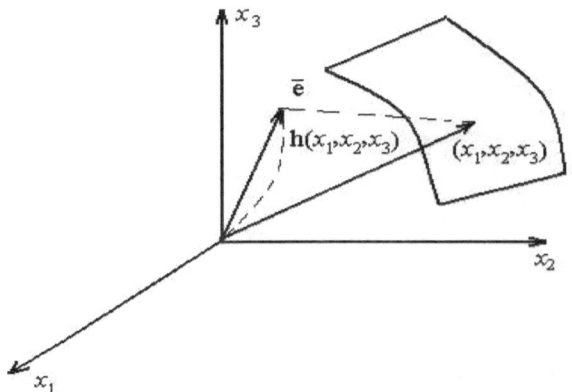

Fig. 2.1.

Fix a sufficient small subdomain $D' \Subset D$ and a locally Lipschitz function φ with a compact support $\operatorname{supp}\varphi \Subset D'$. Consider the vector function

$$f(x,\varepsilon) = (f_1 + \varepsilon\varphi, f_2, \ldots, f_m) : D \to \mathbb{R}^m$$

where $\varepsilon \in (-1,1)$ is an arbitrary number.

This function defines the surface Ω_ε over D. We have

$$\operatorname{area}\Omega_\varepsilon = \iint\limits_{D} \sqrt{g(x,\varepsilon)}\, dx, \quad dx = dx_1\, dx_2\,.$$

Because Ω is locally minimal, we are right to assume that

$$\operatorname{area}\Omega_\varepsilon - \operatorname{area}\Omega = \iint\limits_{D} \sqrt{g(x,\varepsilon)}\, dx - \iint\limits_{D} \sqrt{g(x)}\, dx \geq 0.$$

Here,

$$g(x,\varepsilon) = \det\left(g_{ij}(x,\varepsilon)\right),$$

and

$$g_{ij}(x,\varepsilon) = \left\langle \frac{\partial f}{\partial x_i}(x,\varepsilon), \frac{\partial f}{\partial x_i}(x,\varepsilon) \right\rangle = \left(\frac{\partial f_1}{\partial x_i} + \varepsilon \frac{\partial \varphi}{\partial x_i} \right)\left(\frac{\partial f_1}{\partial x_j} + \varepsilon \frac{\partial \varphi}{\partial x_j} \right) +$$

$$+ \sum_{k=2}^{m} \frac{\partial f_k}{\partial x_i}\frac{\partial f_k}{\partial x_j} = \sum_{k=1}^{m} \frac{\partial f_k}{\partial x_i}\frac{\partial f_k}{\partial x_j} + \varepsilon \left(\frac{\partial \varphi}{\partial x_i}\frac{\partial f_1}{\partial x_j} + \frac{\partial \varphi}{\partial x_j}\frac{\partial f_1}{\partial x_i} \right) +$$

$$+ \varepsilon^2 \frac{\partial \varphi}{\partial x_i}\frac{\partial \varphi}{\partial x_j} = g_{ij} + \varepsilon \left(\frac{\partial \varphi}{\partial x_i}\frac{\partial f_1}{\partial x_j} + \frac{\partial \varphi}{\partial x_j}\frac{\partial f_1}{\partial x_i} \right) + \varepsilon^2 \frac{\partial \varphi}{\partial x_i}\frac{\partial \varphi}{\partial x_j} \,.$$

We prove that

$$\lim_{\varepsilon \to 0} \frac{1}{\varepsilon}\left(\text{area}\,\Omega_\varepsilon - \text{area}\,\Omega \right) = \frac{1}{2} \iint\limits_{D'} \frac{dg}{d\varepsilon}\bigg|_{\varepsilon=0} \frac{dx}{\sqrt{g(x)}} \tag{2.11}$$

where $D' = \text{supp}\,\varphi$.

Since $f,\,\varphi \in \text{Lip}\,D$, then the derivative $\frac{dg}{d\varepsilon}\big|_{\varepsilon=0}$ is finite, and (2.9) guarantees that the right integral (2.11) exists.

We have

$$\frac{1}{\varepsilon}\left(\text{area}\,\Omega_\varepsilon - \text{area}\,\Omega \right) = \frac{1}{\varepsilon} \iint\limits_{D'} \left(\sqrt{g(x,\varepsilon)} - \sqrt{g(x)} \right)\, dx \tag{2.12}$$

$$= \frac{1}{\varepsilon} \iint\limits_{D'} \frac{g(x,\varepsilon) - g(x)}{\sqrt{g(x,\varepsilon)} + \sqrt{g(x)}}\, dx \,.$$

For a.e. $x \in D'$, we are right to assume that

$$g(x,\varepsilon) - g(x) = \varepsilon \frac{dg}{d\varepsilon}\bigg|_{\varepsilon=0} + \int_0^\varepsilon d\eta \int_0^\eta \frac{d^2 g}{d\varepsilon^2}(x,\tau)\, d\tau. \tag{2.13}$$

Decomposing the determinant $g(x,\varepsilon)$ into factors and every time calculating the derivative $dg/d\varepsilon\big|_{\varepsilon=0}$, we see that

$$\frac{dg}{d\varepsilon}\bigg|_{\varepsilon=0} = \frac{1}{2} \sum_{i,j=1}^{2} \left(\frac{\partial \varphi}{\partial x_i}\frac{\partial f_1}{\partial x_j} + \frac{\partial \varphi}{\partial x_j}\frac{\partial f_1}{\partial x_i} \right) \text{adj}\, g_{ij}(x). \tag{2.14}$$

Analogously,

$$\int_0^\varepsilon d\eta \int_0^\eta \frac{d^2 g}{d\varepsilon^2}(x,\tau)\, d\tau = \varepsilon^2\, \alpha(x,\varepsilon) \tag{2.15}$$

where
$$|\alpha(x,\varepsilon)| \leq C(q), \quad \text{for} \quad |\varepsilon| \leq 1,$$

and
$$q = \operatorname*{ess\,sup}_{x \in D'} \max\{|\nabla f_1(x)|, \ldots, |\nabla f_m(x)|, |\nabla \varphi(x)|\}.$$

Now combining (2.13), (2.14) and (2.15), we have

$$\frac{1}{\varepsilon} \iint\limits_{D'} \frac{g(x,\varepsilon) - g(x)}{\sqrt{g(x,\varepsilon)} + \sqrt{g(x)}}\, dx = \frac{1}{2\varepsilon} \iint\limits_{D'} (g(x,\varepsilon) - g(x))\frac{dx}{\sqrt{g(x)}}$$

$$+\frac{1}{\varepsilon} \iint\limits_{D'} (g(x,\varepsilon) - g(x)) \left(\frac{1}{\sqrt{g(x,\varepsilon)} + \sqrt{g(x)}} - \frac{1}{2\sqrt{g(x)}} \right) dx$$

$$= \frac{1}{2} \iint\limits_{D'} \left.\frac{dg}{d\varepsilon}\right|_{\varepsilon=0} \frac{dx}{\sqrt{g(x)}} + \frac{\varepsilon}{2} \iint\limits_{D'} \alpha(x,\varepsilon)\frac{dx}{\sqrt{g(x)}}$$

$$+\frac{1}{2} \iint\limits_{D'} \left(\left.\frac{dg}{d\varepsilon}\right|_{\varepsilon=0} + \varepsilon\alpha(x,\varepsilon) \right) \frac{\sqrt{g(x)} - \sqrt{g(x,\varepsilon)}}{\sqrt{g(x)}\left(\sqrt{g(x,\varepsilon)} + \sqrt{g(x)}\right)}\, dx$$

$$= \frac{1}{2} \iint\limits_{D'} \left.\frac{dg}{d\varepsilon}\right|_{\varepsilon=0} \frac{dx}{\sqrt{g(x)}} + \frac{\varepsilon}{2} \iint\limits_{D'} \alpha(x,\varepsilon)\frac{dx}{\sqrt{g(x)}}$$

$$-\varepsilon \iint\limits_{D'} \left(\frac{1}{2}\left.\frac{dg}{d\varepsilon}\right|_{\varepsilon=0} + \varepsilon\alpha(x,\varepsilon) \right)^2 \frac{dx}{\sqrt{g(x)}\left(\sqrt{g(x,\varepsilon)} + \sqrt{g(x)}\right)}.$$

Since for the last integral it is fulfilled that

$$\left| \iint\limits_{D'} \left(\frac{1}{2}\left.\frac{dg}{d\varepsilon}\right|_{\varepsilon=0} + \varepsilon\alpha(x,\varepsilon) \right)^2 \frac{dx}{\sqrt{g(x)}\left(\sqrt{g(x,\varepsilon)} + \sqrt{g(x)}\right)} \right|$$

$$\leq \frac{1}{8} \iint\limits_{D'} \left(\frac{1}{2}\left.\frac{dg}{d\varepsilon}\right|_{\varepsilon=0} + \varepsilon\alpha(x,\varepsilon) \right)^2 \frac{dx}{g^{\frac{3}{2}}(x)},$$

conditions (2.9), (2.12) imply that (2.11) is true.

Because the surface (2.1) is locally minimal, then the first variation of its area vanishes. In other words, the relation (2.11) implies

$$\iint\limits_{D'} \left.\frac{dg}{d\varepsilon}\right|_{\varepsilon=0} \frac{dx}{\sqrt{g(x)}} = 0.$$

Consequently,

$$\iint_{D'} \sum_{i,j=1}^{2} \left(\frac{\partial \varphi}{\partial x_i} \frac{\partial f_1}{\partial x_j} + \frac{\partial \varphi}{\partial x_j} \frac{\partial f_1}{\partial x_i} \right) \operatorname{adj} g_{ij}(x) \frac{dx}{\sqrt{g(x)}} = 0.$$

Remind that

$$g^{ij} = \frac{\operatorname{adj} g_{ij}}{g}, \quad i, j = 1, 2.$$

Therefore, we can write the last integral in the following form

$$\iint_{D'} \sum_{i,j=1}^{2} \left(\frac{\partial \varphi}{\partial x_i} \frac{\partial f_1}{\partial x_j} + \frac{\partial \varphi}{\partial x_j} \frac{\partial f_1}{\partial x_i} \right) g^{ij}(x) \sqrt{g(x)} \, dx = 0,$$

or

$$\iint_{D'} \sum_{i=1}^{2} \frac{\partial \varphi}{\partial x_i} \sum_{j=1}^{2} g^{ij}(x) \frac{\partial f_1}{\partial x_j} \sqrt{g(x)} \, dx = 0.$$

Thus the function $f_1(x)$ is a generalized solution of (2.4). $\qquad \square$

Theorem 2.16 *If a locally Lipschitz surface Ω in $\mathbb{R}^m$ given in the nonparametric form*

$$\xi = f(x_1, x_2) = (x_1, x_2, f_1(x), \ldots, f_{m-2}(x)) : D \to \mathbb{R}^m \qquad (2.17)$$

is locally minimal, then the vector function f satisfies (in the generalized sense) to the system of partial differential equations

$$\sum_{i=1}^{2} \frac{\partial}{\partial x_i} \left(\sqrt{g} \sum_{j=1}^{2} g^{ij}(x) \frac{\partial f_k}{\partial x_j} \right) = 0, \quad 1 \le k \le m - 2. \qquad (2.18)$$

For the **proof** it is sufficient to observe that in the case of the nonparametric representation (2.17) for the coefficients of the first quadratic form of Ω, it is fulfilled

$$g_{11} = 1 + \sum_{k=1}^{m-2} \left(\frac{\partial f_k}{\partial x_1} \right)^2, \quad g_{12} = \sum_{k=1}^{m-2} \frac{\partial f_k}{\partial x_1} \frac{\partial f_k}{\partial x_2}, \quad g_{22} = 1 + \sum_{k=1}^{m-2} \left(\frac{\partial f_k}{\partial x_2} \right)^2.$$

Since

$$g = g_{11} g_{22} - g_{12}^2 \ge 1 + \sum_{k=1}^{m-2} \left(\frac{\partial f_k}{\partial x_1} \right)^2 + \sum_{k=1}^{m-2} \left(\frac{\partial f_k}{\partial x_2} \right)^2 \ge 1,$$

then the assumption (2.9) is fulfilled always. Theorem 2.10 implies that the system (2.18) is valid. $\qquad \square$

Corollary 2.19 *If a surface Ω in $\mathbb{R}^3$ is given in the form of a graph of a locally Lipschitz function $x_3 = f(x_1, x_2)$ and locally minimal, then f is a generalized solution of the equation*

$$\sum_{i=1}^{2} \frac{\partial}{\partial x_i} \left(\frac{f'_{x_i}}{\sqrt{1 + |\nabla f|^2}} \right) = 0. \tag{2.20}$$

Open questions 2.21 1) Is condition (2.9) essential for the validity of Theorem 2.10 ? 2) Find conditions, allowing to replace (2.9) by a weaker condition. 3) Find analogs of Theorem 2.10 for non-regular surfaces which belong to wider classes than the class of locally Lipschitz surfaces. 4) Find an analog of Theorem 2.10 for non-regular surfaces in pseudo-Euclidean spaces (A.N. Kondrashow).

Chapter 3

Isothermal Coordinates

Below we prove the existence of isothermal coordinates on non-regular surfaces of some special classes and research smooth properties of isothermal representations. We follow [118].

3.1   Main Theorem

Let $D \subset \mathbb{R}^2$ be a domain and let Ω be a two-dimensional surface in $\mathbb{R}^m$, $m \geq 3$, given by a locally Lipschitz vector function (2.1). We assume that a.e. on D

$$\mathrm{rank}\,(df) = 2\,. \tag{3.1}$$

The assumption (3.1) is equivalent to the condition $g = g_{11}g_{22} - g_{12}^2 > 0$ a.e. on D.

Exercise 3.2   If a surface Ω is locally bi-Lipschitz, then (3.1) is valid at each point where f is differentiable (prove or refute).

$\square$

At each point $x \in D$ where a vector function f is differentiable and satisfies (3.1), (2.3), the mapping $f : D \to \Omega$ conserves angles between curves, and it is conformal in the traditional sense.

We will need the conception of $W^{1,2}$-majorized functions.

Definition 3.3   *Let $D \subset \mathbb{R}^2$ be a domain. We say that a function $P : D \to \mathbb{R}$ is $W^{1,2}$-majorized on D if there exists a function $K \in W^{1,2}(D)$ such that*

$$P(x) \leq K(x) \quad \text{for a.e.} \quad x \in D\,. \tag{3.4}$$

42

A function P is called locally $W^{1,2}$-majorized on D if it is $W^{1,2}(D')$-majorized for every subdomain $D' \Subset D$.

The bounded functions are simplest examples of $W^{1,2}$-majorized functions. Let $P : D \to \mathbb{R}$ be a bounded function defined over a domain D with finite area. Here we are able to choose $K = \sup_{x \in D} P(x)$. It is clear that $K \in W^{1,2}(D)$, and (3.4) holds everywhere on D.

In the general case, the volume of the class of $W^{1,2}$-majorized functions is not explicit. See the discussion on this question at the end of Chapter 5.

Let D be a simply connected domain in $\mathbb{R}^2$ and let Ω be a surface, given over D by a vector function (2.1), satisfying (3.1). Let $x = (x_1, x_2) \in D$ be a point at which where the vector function f is differentiable and (3.1) holds. The assumption (3.1) implies that $df \neq 0$ at this point the metric ds_Ω is non-degenerate and a infinitesimal circle with respect to the metric ds_Ω with the center x is an infinitesimal ellipse with respect to the Euclidean metric. Denote by $p(x) \geq 1$ and $0 \leq \theta(x) < \pi$ characteristics of this ellipse, i.e. the ratio p the biggest axis of the ellipse to the simplest axis and the angle θ between the biggest axis and the direction $\overrightarrow{Ox_1}$.

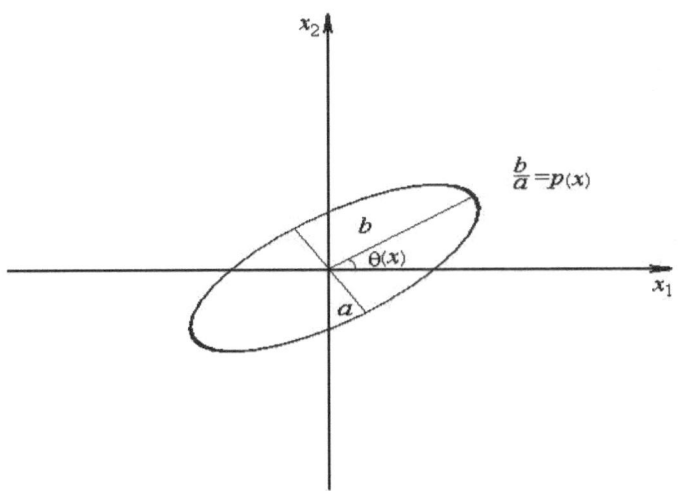

Fig. 3.1.

It is easy to see that

$$p(x) = \frac{g_{11}(x) + g_{22}(x)}{2\sqrt{g_{11}(x)g_{22}(x) - g_{12}^2(x)}} + \left(\frac{(g_{11}(x) + g_{22}(x))^2}{4\left(g_{11}(x)g_{22}(x) - g_{12}^2(x)\right)} - 1 \right)^{1/2}.$$

The characteristic $\theta(x)$ is defined at each point where $p(x) > 1$.

We recall some necessary concepts on $W^{1,2}$-functions. Let D be a domain in $\mathbb{R}^2$. A function $f \in L^p(D)$, $p \geq 1$, belongs to a class $W^{1,p}(D)$ if f is absolutely continuous along a.e. vertical and horizontal cross-sections of D and its partial derivatives belong to $L^p(D)$. A function $f \in W_{loc}^{1,p}(D)$ if $f \in W^{1,p}(D_1)$ for every open subset $D_1 \Subset D$ (see, for example, Goldstein, Reshetnyak [46, Chapter 2, 5.6]).

Observe also (see Gehring, Lehto [42] or Zhukov [202]) that a function $f : D \to \mathbb{R}$ is differentiable a.e. on D if $f \in W_{loc}^{1,p}(D)$ for $p > 2$ or $f \in W_{loc}^{1,1}(D)$ and is monotone in the Lebesgue sense, i.e.

$$\operatorname{osc}(D_1, f) \leq \operatorname{osc}(\partial D_1, f) \quad \text{for every subdomain} \quad D_1 \Subset D.$$

A vector function $f = (f_1, f_2, \ldots, f_m) : D \to \mathbb{R}^m$ is monotone if every f_i is monotone, $i = 1, 2, \ldots, m$.

Every monotone, in the Lebesgue sense, vector function $f : D \to \mathbb{R}^m$ of the class $W_{loc}^{1,2}(D)$ is differentiable a.e. on D. In particular, a homeomorphic map $f : D \to \mathbb{R}^2$ of the class $W_{loc}^{1,2}(D)$ is differentiable a.e. on D.

Definition 3.5 *A homeomorphic mapping $f : D \to \mathbb{R}^2$ of the class $W_{loc}^{1,2}$ on a domain $D \subset \mathbb{R}^2$ is called quasi-conformal with M.A. Lavrentiev characteristics $(p(x), \theta(x))$ if a.e. on D it transforms infinitesimal ellipses with characteristics $(p(x), \theta(x))$ into infinitesimal circles[1]. A mapping $f : D \to \mathbb{R}^2$ is called q-quasi-conformal if* ess sup$_{x \in D}$ $p(x) \leq q$.

On quasi-conformal maps of two-dimensional domains, see Volkovyskii [192], Lehto and Virtanen [90], Belinskii [17].

We fix a number sequence $\{Q_n\}_{n=1}^\infty$ with properties

$$Q_n \geq 1 \quad \text{for all} \quad n = 1, 2, \ldots \quad \text{and} \quad Q_n \to \infty \quad \text{for} \quad n \to \infty.$$

For every $n = 1, 2, \ldots$ we put

$$p_n(x) = \begin{cases} p(x), & \text{for } p(x) \leq Q_n, \\ 1, & \text{for } p(x) > Q_n. \end{cases}$$

By well-known results of the quasi-conformal maps theory (see, for example, Belinskii [17, Theorem 8]), there exists a quasi-conformal mapping $\xi = w_n(x)$

---

[1]This means that for a.e. $x \in D$ the differential $df(x) : \mathbb{R}^2 \to \mathbb{R}^2$ has characteristics $(p(x), \theta(x))$.

of a domain D onto a suitable domain $\Delta_n \subset \mathbb{R}^2$ with characteristics coinciding with (p_n, θ) a.e. on D. This domain Δ_n can be either the same as $\mathbb{R}^2$ if $D = \mathbb{R}^2$, or an arbitrary proper simply connected subdomain of $\mathbb{R}^2$.

Here and below, by $B(a, r)$ we denote a disc of a radius $r > 0$ with a center at $a \in \mathbb{R}^2$, and by $S(a, r)$ we denote its boundary.

We fix two points a_0, $a_1 \in D$. By auxiliary conformal transformations on the plane of the variable $\xi = (\xi_1, \xi_2)$, we attain that the domains Δ_n are discs $B_n = B(0, R_n)$ with $1 < R_n < \infty$, and the maps w_n satisfy the following conditions:

$$w_n(a_0) = (0,0) \quad \text{and} \quad w_n(a_1) = (1,0) \quad (n = 1, 2, \ldots). \tag{3.6}$$

The sequence of maps $F_n = f \circ w_n^{-1} : B_n \to \mathbb{R}^m$ where $w_n : D \to B_n$ satisfy (3.6), we call the *canonical sequences of representations* of the surface Ω, corresponding with the number sequence $\{Q_n\}$.

Fix a canonical sequence of representations $F_n : B_n \to \mathbb{R}^m$ of Ω.

Definition 3.7 *We will say that a sequence $\{F_n\}$ converges locally uniform to a canonical homeomorphism $F : B(0, R) \to \mathbb{R}^m$ if:*

(i)   the domains B_n converge as $n \to \infty$ to a domain $B = B(0, R)$ as a kernel with respect to the point $\xi = 0$ (in other words, there exists $R = \lim\limits_{n\to\infty} R_n$, $1 \le R \le \infty)^2$;

(ii)   the sequence $\{F_n\}$ converges as $n \to \infty$ to $F : B(0, R) \to \mathbb{R}^m$ uniformly on every subdomain $U \subset\subset B(0, R)$;

(iii)   the sequence of inverse mappings $F_n^{-1} : \Omega \to \mathbb{R}^2$ converges as $n \to \infty$ to the mapping $F^{-1} : \Omega \to B(0, R)$ uniformly on every subdomain $\Omega_1 \subset\subset \Omega$.

The main result of this chapter provides the following existence theorem [118].

Theorem 3.8 *Let Ω be a two-dimensional surface, given by a vector function (2.1) over a bounded simply connected domain $D \subset \mathbb{R}^2$ and satisfying to (3.1). Suppose that a function $\mathcal{P}$ defined by*

$$\mathcal{P}(x) = (g_{11}(x) + g_{22}(x)) \Big/ 2\sqrt{g_{11}(x)\, g_{22}(x) - g_{12}^2(x)} \tag{3.9}$$

is $W^{1,2}$-majorized on D.

---

[2]See definition of the Caratéodory kernel convergence in [177, Part II], [159], Nasyrov [136].

Then there exist on Ω isothermal coordinates $\xi = (\xi_1, \xi_2)$, introduced by a canonical homeomorphism $y = F(\xi) : B(0, R) \subset \mathbb{R}^2 \to \Omega$, limiting for a canonical representations sequence $F_n(\xi) : B(0, R_n) \to \Omega$ of Ω, corresponding to a preassigned number sequence $\{Q_n\}$. Moreover, there exists $\lim_{n\to\infty} R_n = R$, $1 < R < \infty$, and the sequence $\{F_n\}$ converges to $F(\xi)$ locally uniformly on $B = B(0, R)$.

The canonical homeomorphism $F : B \to \Omega$ is unique.

During the proof of Theorem 3.8, we obtain some concrete two-sides estimates of the radius R with the distance $|a_1 - a_0|$, the integral

$$\iint\limits_{D} K(x) \, dx_1 dx_2$$

and the infimum of diameters of arcs with the ends on ∂D separating the point a_1 from the point a_0 (inequalities (3.34) and (3.37) in §5).

Differential properties of the canonical homeomorphism

$$F : B(0, R) \to \mathbb{R}^m \,,$$

and its conversion are defined by properties f and f^{-1}. They have been researched separately. Chapter 5 describes some estimates of an approximation speed F by F_n.

3.2   Canonical Homeomorphisms

Let Ω be a surface of the form (2.1). Since the metric on $f(D)$ is induced from $\mathbb{R}^m$, then on Ω it is defined the 2-dimensional Hausdorff measure $\mathrm{mes}_2(E)$, $E \subset f(D)$.[3] If $\Omega_1 \subset \mathbb{R}^n$ and $\Omega_2 \subset \mathbb{R}^m$ are two-dimensional surfaces and $h : \Omega_1 \to \Omega_2$ is a homeomorphism, then we say that h has the Lusin N-property if for every set $E \subset \Omega_1$, $\mathrm{mes}_2(E) = 0$, it is fulfilled $\mathrm{mes}_2(h(E)) = 0$ (Saks [161, Chapter VII, §6]).

Suppose that Ω is a surface of the form (2.1) satisfying (3.1). Suppose also that a vector function $f \in W^{1,2}_{\mathrm{loc}}(D)$. Since the map $f : D \to f(D)$ is homeomorphic, then for every set $E \subset D$ it is fulfilled

$$\mathrm{mes}_2(f(E)) = \iint\limits_{E} \sqrt{g_{11}g_{22} - g_{12}^2} \, dx_1 dx_2 \qquad (3.10)$$

---

[3] Sometimes we will use also another standard designation $\mathrm{mes}_2(E) = \mathrm{area}(E)$ or for brevity $|E|$.

(see Hajlasz [56, Theorem **12**]).

From (3.1) and (3.10) it follows that

$$\text{mes}_2\, (f(E)) = 0 \quad \text{if and only if} \quad \text{mes}_2\, (E) = 0 \qquad (3.11)$$

and the maps f, f^{-1} have the Lusin N-property.

Because f is a.e. differentiable, then by (3.11) a.e. on Ω there exists a tangent plane $T_y(\Omega)$. Thus a.e. on Ω, we can define a gradient (connection) ∇_Ω.

Indeed, let $\varphi : \Omega \to \mathbb{R}$ be a function and let $y \in \Omega$ be a fixed point in which the tangent plane $T_y(\Omega)$ exists. Following the standard scheme into the introduction of the connection, we extend φ onto a neighborhood of y in $\mathbb{R}^m$. Let $\tilde{\varphi}$ be this new function. By the definition we put

$$\nabla_\Omega \varphi(y) = \left(\overline{\nabla} \tilde{\varphi}(y)\right)^T .$$

Here $\overline{\nabla} = \nabla_{\mathbb{R}^m}$ is the standard differentiation in $\mathbb{R}^m$ and the symbol u^T means the orthogonal projection of a vector u onto the tangent plane $T_y(\Omega)$.

Let $U \Subset \Omega$ be an open boundary set. By $W^{1,p}(U)$, $p \geq 1$, we denote the set of all functions $\varphi : U \to \mathbb{R}$ with the following property: *for every point $x \in U$, there exists a neighborhood $V \subset \mathbb{R}^m$ and a sequence of C^1-functions $\varphi_k : V \to \mathbb{R}$ such that the contractions $\tilde{\varphi}_k = \varphi_k|_{V \cap U}$ converge to φ uniformly; moreover,*

$$\sup_{k \geq 1} \int_{V \cap U} |\nabla_\Omega \tilde{\varphi}_k|^p \, d\,\text{area}_\Omega \leq q < \infty$$

for some $q = \text{const} < \infty$. By $W^{1,p}_{\text{loc}}(\Omega)$ we denote the set of functions $\varphi : \Omega \to \mathbb{R}$ which belong to $W^{1,p}(U)$ for every open subset $U \Subset \Omega$.

By Theorem 3.8 a canonical homeomorphism $F : B(0,R) \to \Omega$ is differentiable a.e. on the domain $B(0,R)$. If a vector function $f \in C^k(D)$, $k \geq 1$, then the map F defines isothermal coordinates of a class C^k on Ω and it is a conformal map of a plane domain D onto Ω in the usual sense.

Conformal maps of nonregular surfaces were studied only in very special occasions. Schwarz proved that surface of every tetrahedron and cube may be mapped onto the unit sphere (see Carathéodory [25, §**161**]). On conformal maps of general polyhedral surfaces see Gresky [49], Ells and Fuglede [33] and of manifolds of bounded curvature see Reshetnyak [149] and also Toro [185], Müller and Sverák [134]. A generalized Cristoffel-Schwartz formula for piecewise conformal homeomorphisms and close questions were considered by Grudskii [53], [54].

In the considered case of $W^{1,2}_{\text{loc}}$-surfaces we may say only on conformality at points where f is differentiable, i.e. on conformality of $f : D \to \Omega$ a.e. on D.

What is 'conformality in the large' here is not clear, because the definition of conformal mappings of nonregular surfaces meets essential difficulties like, in particular, with the well-known in the plane case problem of the removability of singular sets. As in the plane case we can study different classes of mappings conformal outside of the singular set and having different properties 'in the large' (see, for example, Menchoff [105], Trohimchuk [186], Dolgenko [31], Havinson [58], Kühnau [80], David and Mattila [30], others).

The following theorem establishes some connections between properties of $f : D \to \mathbb{R}^m$ and properties of the canonical homeomorphism F.

Theorem 3.12   *Let Ω be a surface given by a vector – function (2.1) over a simply connected bounded domain $D \subset \mathbb{R}^2$ and satisfying (3.1). Suppose that a function $\mathcal{P}$ defined by (3.9) is $W^{1,2}$-majorized on D. Then we have:*

(i)   the canonical homeomorphism $F : B(0, R) \to \Omega$ and the inverse homeomorphism $F^{-1} : \Omega \to B(0, R))$ possess the Lusin N-property if and only if $f : D \to \Omega$ (respectively, $f^{-1} : \Omega \to D$) possess the N-property;

(ii)   if $f \in W^{1,2}_{\mathrm{loc}}(D)$, then the canonical homeomorphism $F : B(0, R) \to \Omega$ belongs to $W^{1,1}_{\mathrm{loc}}(B(0, R))$;

(iii)   if $f^{-1} \in W^{1,\alpha}_{\mathrm{loc}}(\Omega)$ for a number α, $2 \le \alpha \le \infty$, and there exists β, $1/\alpha + 1/\beta = 1/2$, such that for every subdomain $D_1 \Subset D$

$$\iint\limits_{D_1} \left(g_{11}g_{22} - g_{12}^2\right)^{\beta/2} dx_1 dx_2 < \infty, \qquad (3.13)$$

then the inverse map $F^{-1} : \Omega \to B(0, R)$ belongs to $W^{1,1}_{\mathrm{loc}}(\Omega)$.

The proofs of these statements will be given below. We formulate beforehand some their corollaries for nonparametric surfaces.

3.3   Nonparametric Surfaces

Let D be a bounded simply connected subdomain of $\mathbb{R}^2$ and let Ω be a two-dimensional surface in $\mathbb{R}^m$ given by a system of functions:

$$
\begin{aligned}
y_1 &= x_1, \\
y_2 &= x_2, \\
y_3 &= f_1(x_1, x_2), \\
&\ldots\ldots \\
y_m &= f_{m-2}(x_1, x_2)
\end{aligned}
\qquad (3.14)
$$

where $x = (x_1, x_2) \in D$.

The system (3.14) is a special case (2.1). Here we have

$$g_{11}(x) = 1 + \sum_{i=1}^{m-2} \left| \frac{\partial f_i}{\partial x_1}(x) \right|^2 ,$$

$$g_{12}(x) = \sum_{i=1}^{m-2} \frac{\partial f_i}{\partial x_1}(x) \frac{\partial f_i}{\partial x_2}(x) ,$$

$$g_{22}(x) = 1 + \sum_{i=1}^{m-2} \left| \frac{\partial f_i}{\partial x_2}(x) \right|^2 .$$

Now we obtain

$$g_{11}(x)g_{22}(x) - g_{12}^2(x) = 1 + \sum_{i=1}^{m-2} |\nabla f_i(x)|^2 +$$

$$+ \sum_{i=1}^{m-2} \left| \frac{\partial f_i}{\partial x_1}(x) \right|^2 \sum_{i=1}^{m-2} \left| \frac{\partial f_i}{\partial x_2}(x) \right|^2 - \left(\sum_{i=1}^{m-2} \frac{\partial f_i}{\partial x_1}(x) \frac{\partial f_i}{\partial x_2}(x) \right)^2 \geq$$

$$\geq 1 + \sum_{i=1}^{m-2} |\nabla f_i(x)|^2 .$$

Suppose that f_i, $i = 1, \ldots, m-2$, are Lipschitz functions on D. There exists a constant $0 < M < \infty$ such that

$$|\nabla f| = \left(\sum_{i=1}^{m-2} |\nabla f_i(x)|^2 \right)^{1/2} \leq M \quad \text{a.e. on} \quad D.$$

For the function $\mathcal{P}$ defined by (3.9) a.e. on D we have

$$\mathcal{P}(x) \leq \left(1 + \sum_{i=1}^{m-2} |\nabla f_i(x)|^2 \right)^{1/2} . \tag{3.15}$$

Thus $\mathcal{P}$ is bounded and satisfies (3.4).

The statements on the existence of isothermal coordinates on nonparametric surfaces in $\mathbb{R}^m$ formulated below are simple combinations of theorems 3.8 and 3.12.

Corollary 3.16   *Let $D \subset \mathbb{R}^2$ be a bounded simply connected domain. If a nonparametric surface (3.14) satisfies to the Lipschitz condition, then there exist isothermal coordinates $\xi = (\xi_1, \xi_2)$ given by a canonic homeomorphism $y = F(\xi) : B(0, R) \to \Omega$.*

Moreover, the disc $B(0, R)$ has a radius $1 < R < \infty$ and the maps F, $F^{-1} \in W^{1,2}_{\mathrm{loc}}$ have the Lusin N-property.

The canonical homeomorphism F is unique.

For surfaces of the classes $W^{l,p}$ with $l \geq 3$ and $p > 2$, this statement can be obtained from well-known results of Bojarski and Vekua (see [189, Chapter 2, §6]). A close statement for Lipschitz graphs belongs to Morrey [133], and for locally Lipschitz graphs it follows from results of Bers [21], Belinskii [17, Theorem **9**], Martio and Miklyukov [98].

Corollary 3.17   *Let Ω be a two-dimensional non-parametric surface given by a vector-function (3.14) over a bounded simply connected domain $D \subset \mathbb{R}^2$ where $f_i \in W^{1,2}(D)$ $(i = 1, \ldots, m - 2)$. If the function $|\nabla f|$ is $W^{1,2}$-majorized on D, then there exist isothermal coordinates $\xi = (\xi_1, \xi_2)$ defined by a canonical homeomorphism $y = F(\xi) : B(0, R) \subset \mathbb{R}^2 \to \Omega$.*

Moreover, the disc $B(0, R)$ has the radius $1 < R < \infty$, the mappings F, F^{-1} have the Lusin N-property and such that $F \in W^{1,2}(B(0, R))$, $F^{-1} \in W^{1,2}_{\mathrm{loc}}(\Omega)$.

The canonical homeomorphism F is unique.

If a nonparametric surface Ω is given over a domain D by a vector function (3.14) where every function

$$f_i, \quad i = 1, 2, \ldots, m - 2,$$

belongs to $W^{2,2}(D)$ then, obviously, the function $|\nabla f|$ is $W^{1,2}$-majorized. Thus *it is possible to introduce isothermal coordinates on every two-dimensional nonparametric surface in $\mathbb{R}^m$ of the class $W^{2,2}(D)$.*

Example 3.18   We will illustrate these results by a simple example showing applicability of them to surfaces with cusps. Consider the graph of the function

$$y_3 = \int\limits_0^r \ln^\varepsilon \frac{1}{s}\, ds, \quad r = \sqrt{y_1^2 + y_2^2}, \quad (y_1, y_2, y_3) \in \mathbb{R}^3.$$

If $0 < \varepsilon < \frac{1}{2}$, then this function belongs to $W^{2,2}(B)$ on the disc $B = B(0, 1)$. Its graph has a cusp at the origin of $\mathbb{R}^3$.

Using this function, it is not difficult to construct examples of surfaces, satisfying assumptions of Corollary 3.17 and having sufficiently massive sets of cusps (in the sense of the Hausdorff measure).

3.4 Proof of Theorem 3.12

Fix a canonical homeomorphism $F : B(0, R) \to \Omega$ and a canonical sequence of representations

$$F_n : B_n \to \Omega, \quad F_n = f \circ w_n^{-1}$$

described by Theorem. Every map $w_n : D \to B_n$ is Q_n-quasi-conformal and by the boundedness of the domains D and B_n, the map w_n and the inverse map w_n^{-1} belong to classes $W^{1,2}$ on corresponding domains.

Here and below, for a map $u(x) = (u_1, u_2) : D \to \mathbb{R}^2$ of the class $W^{1,1}_{loc}(D)$ at every point $x = (x_1, x_2) \in D$ where u is differentiable, we put

$$\Lambda(x, u) = |\nabla u_1(x)|^2 + |\nabla u_2(x)|^2.$$

By $J(x, u)$ we denote the Jacobian

$$J(x, u) = (u_1(x))'_{x_1} (u_2(x))'_{x_2} - (u_1(x))'_{x_2} (u_2(x))'_{x_1}.$$

The key role in the proof is the following a priory estimate of the Dirichlét integral for quasi-conformal mappings $w_n : D \to \mathbb{R}^2$.

Lemma 3.19 *Let $U \Subset D$ be a subdomain of a domain D. Suppose that the function $\mathcal{P}(x)$ defined by (3.9) is $W^{1,2}$-majorized on D so that for a function $K \in W^{1,2}(D)$ a.e. on D it is fulfilled $\mathcal{P}(x) \le K(x)$.*

Then the following estimate holds

$$\int_U \Lambda(x, w_n)\, dx_1 dx_2 \le c_1 \int_D |w_n|^2 \left(K^2 + |\nabla K|^2 \right) dx_1 dx_2 \qquad (3.20)$$

where constant c_1 depends on U and D only.

The idea of the *proof* belongs to Miklyukov and Suvorov [125]. As we will see from below, we are right to assume that every quasi-conformal map $w_n : D \to \mathbb{R}^2$, $n = 1, 2, \dots$ belongs to the class $C^2(D)$. It is possible by means of a suitable approximation of these maps by smooth mappings.

Let η, ζ and h be C^2-functions defined on a domain $V \subset \mathbb{R}^2$ with a smooth boundary ∂V. By Green's formula we have

$$\int_{\partial V} \eta\, \zeta\, dh = \iint_V d(\eta\, \zeta) \wedge dh =$$

$$= \iint_V \eta\, d\zeta \wedge dh + \iint_V \zeta\, d\eta \wedge dh$$

where the symbol $\wedge$ means the exterior product of differential 1-forms.

If η vanishes on ∂V, then we arrive to the following formula of 'part integration'

$$\iint_V \eta \, d\zeta \wedge dh = - \iint_V \zeta \, d\eta \wedge dh. \qquad (3.21)$$

Now we fix a sequence $\{K_l\}$ $(l = 1, 2, \ldots)$ of C^1-functions with the property

$$\|K_l - K\|_{W^{1,2}(D)} \to 0 \quad \text{for} \quad l \to \infty. \qquad (3.22)$$

Choose a nonnegative C^1-function $\varphi : D \to \mathbb{R}$ with a compact support $\operatorname{supp} \varphi \subset D$ such that $\varphi \equiv 1$ on U.

In (3.21) we put

$$\eta = \varphi^2 K_l, \quad \zeta = u_{1n}, \quad h = u_{2n}$$

and let

$$V \quad \text{be the interior of the compact} \quad \operatorname{supp} \varphi.$$

Without loss of generality we can assume that the set V is a connected domain with a smooth boundary. By (3.21) we have

$$\iint_V \varphi^2 K_l J(x, w_n) \, dx_1 dx_2 = - \iint_V u_{1n} \, d\left(\varphi^2 K_l\right) \wedge du_{2n} =$$

$$= -2 \iint_V \varphi \, u_{1n} K_l \, d\varphi \wedge du_{2n} - \iint_V \varphi^2 \, u_{1n} \, dK_l \wedge du_{2n}.$$

For every $l = 1, 2, \ldots$ we obtain

$$\iint_V \varphi^2 K_l J(x, w_n) \, dx_1 dx_2 =$$

$$= -2 \iint_V \varphi \, u_{1n} K_l \left(\varphi'_{x_1} (u_{2n})'_{x_2} - \varphi'_{x_2} (u_{2n})'_{x_1}\right) dx_1 dx_2 -$$

$$- \iint_V \varphi^2 u_{1n} \left(K'_{lx_1} (u_{2n})'_{x_2} - K'_{lx_2} (u_{2n})'_{x_1}\right) dx_1 dx_2.$$

This implies to the estimate

$$\iint\limits_{D} \varphi^2 K_l \, J(x, w_n) \, dx_1 dx_2 \le$$

$$\le 2 \iint\limits_{D} \varphi \, |u_{1n}| \, |K_l| \, \left| \varphi'_{x_1} (u_{2n})'_{x_2} - \varphi'_{x_2} (u_{2n})'_{x_1} \right| \, dx_1 dx_2 +$$

$$+ \iint\limits_{D} \varphi^2 |u_{1n}| \, \left| (K_l)'_{x_1} (u_{2n})'_{x_2} - (K_l)'_{x_2} (u_{2n})'_{x_1} \right| \, dx_1 dx_2$$

whence it follows that

$$\iint\limits_{D} \varphi^2 \, K_l \, J(x, w_n) \, dx_1 dx_2 \le 2 \iint\limits_{D} \varphi \, |u_{1n}| \, |K_l| \, |\nabla \varphi| \, |\nabla u_{2n}| \, dx_1 dx_2 +$$

$$+ \iint\limits_{D} \varphi^2 \, |u_{1n}| \, |\nabla K_l| \, |\nabla u_{2n}| \, dx_1 dx_2 \, .$$

Now setting $l \to \infty$ and using (3.22), we arrive to the following estimate

$$\iint\limits_{D} \varphi^2 \, K \, J(x, w_n) \, dx_1 dx_2 \le 2 \iint\limits_{D} \varphi \, |u_{1n}| K \, |\nabla \varphi| \, |\nabla u_{2n}| \, dx_1 dx_2 +$$

$$+ \iint\limits_{D} \varphi^2 \, |u_{1n}| \, |\nabla K| \, |\nabla u_{2n}| \, dx_1 dx_2 \, .$$

$$(3.23)$$

Next we observe that from the definition of quasi-conformality, it follows

$$|\nabla u_{1n}|^2 + |\nabla u_{2n}|^2 \ = (p_n + 1/p_n) \, J(x, w_n) \le$$

$$\le 2 p_n \, J(x, w_n) \le 2K(x) \, J(x, w_n) \, . \qquad (3.24)$$

Combining (3.23) and the 'Cauchy inequality with $\epsilon > 0$'

$$|ab| \le \frac{\epsilon^2}{2} a^2 + \frac{1}{2\epsilon^2} b^2 \, ,$$

we have

$$\iint\limits_{D} \varphi^2 \left(|\nabla u_{1n}|^2 + |\nabla u_{2n}|^2 \right) dx_1 dx_2 \le$$

$$\le 2 \iint\limits_{D} \varphi^2 K(x) J(x, w_n) dx_1 dx_2 \le$$

$$\le 4 \iint\limits_{D} \varphi |u_{1n}| K| \nabla \varphi| |\nabla u_{2n}| dx_1 dx_2 +$$

$$+2 \iint\limits_{D} \varphi^2 |u_{1n}| |\nabla K| |\nabla u_{2n}| dx_1 dx_2 \le$$

$$\le 3\epsilon^2 \iint\limits_{D} \varphi^2 |\nabla u_{2n}|^2 dx_1 dx_2 + \frac{2}{\epsilon^2} \iint\limits_{D} |u_{1n}|^2 K^2 |\nabla \varphi|^2 dx_1 dx_2 +$$

$$+\frac{1}{\epsilon^2} \iint\limits_{D} \varphi^2 |u_{1n}|^2 |\nabla K|^2 dx_1 dx_2.$$

Choose $\epsilon > 0$ so that $3\epsilon^2 < 1$ (for example, $\epsilon = 1/2$.) Then

$$\iint\limits_{D} \varphi^2 \left(|\nabla u_{1n}|^2 + |\nabla u_{2n}|^2 \right) dx_1 dx_2 \le$$

$$\le c_2 \left(\iint\limits_{D} |u_{1n}|^2 K^2 |\nabla \varphi|^2 dx_1 dx_2 + \iint\limits_{D} \varphi^2 |u_{1n}|^2 |\nabla K|^2 dx_1 dx_2 \right)$$

where c_2 is a constant.

Since $\varphi \equiv 1$ on U, we obtain

$$\iint\limits_{U} \Lambda(x, w_n) dx_1 dx_2 \le$$

$$\le c_2 \left(M_1(\varphi) \iint\limits_{D} |w_n|^2 K^2 dx_1 dx_2 + M_2(\varphi) \iint\limits_{D} |w_n|^2 |\nabla K|^2 dx_1 dx_2 \right).$$

Here the constants

$$M_1(\varphi) = \sup_{x \in D} |\nabla \varphi(x)|^2, \quad M_2(\varphi) = \sup_{x \in D} \varphi^2(x)$$

can be made depending on domains U and D only. This remark completes the proof of Lemma 3.19. □

We need an estimate of the Dirichlét integral for inverse maps $w_n^{-1} : B_n \to D$ which is valid for every fixed $n = 1, 2, \ldots$.

Lemma 3.25 *Let $U \subset D$ and $V \subset B_n$ be subdomains and $w_n(U) \supset V$. If the characteristic $p(x)$ is $W^{1,2}$-majorized on D, then*

$$\iint_V \Lambda(\xi, w_n^{-1}) \, d\xi_1 d\xi_2 \leq \iint_U K(x) \, dx_1 dx_2. \tag{3.26}$$

Proof is simple. Recall the formula of the change of variables for $W^{1,2}$-homeomorphisms $w_n : D \to B_n$ (see, for example, [46, Theorem **2.4**]). For a Lebesgue measurable function $h(\xi)$ and every $n = 1, 2, \ldots$, we have

$$\iint_{B_n} h(\xi) \, d\xi_1 d\xi_2 = \iint_D h \circ w_n \, J(x, w_n) \, dx_1 dx_2. \tag{3.27}$$

Let $\xi \in B_n$ be a point at which w_n^{-1} is differentiable and $dw_n^{-1}(\xi) \neq 0$. The infinitesimal disc with the center at $w_n^{-1}(\xi)$ has an infinitesimal ellipse as its preimage under dw_n^{-1}. Denote by $p(\xi, w_n^{-1})$ the ratio of the larger axis of this ellipse to the smaller axis. If $\xi = w_n(x)$, then, obviously,

$$p(\xi, w_n^{-1}) = p_n(x) \leq K(x). \tag{3.28}$$

Since maps w_n and w_n^{-1} are quasi-conformal, then they have the Lusin N-property, and (3.28) holds for a.e. $x \in D$ and a.e. $\xi \in B_n$.

Using (3.24), we have

$$\iint_V \Lambda(\xi, w_n^{-1}) \, d\xi_1 d\xi_2 \leq 2 \iint_V p(\xi, w_n^{-1}) \, J(\xi, w_n^{-1}) \, d\xi_1 d\xi_2.$$

By (3.27) and (3.28), we obtain

$$\iint_V p(\xi, w_n^{-1}) \, J(\xi, w_n^{-1}) \, d\xi_1 d\xi_2 = \iint_{w_n^{-1}(V)} p_n(x) \, dx_1 dx_2.$$

The assumption $w_n(U) \supset V$ with (3.28) implies to the necessary estimate. □

Return to the proof of Theorem 3.12. By the definition, the sequence F_n^{-1} : $\Omega \to B_n$, $n = 1, 2, \ldots$ converges to the homeomorphism $F^{-1} : \Omega \to B(0, R)$ uniformly on every subdomain $\Omega_1 \subset\subset \Omega$. Since $F_n^{-1} = w_n \circ f^{-1}$, then the sequence of Q_n-quasi-conformal maps $w_n = F_n^{-1} \circ f$ satisfies (3.6) and converges to the homeomorphism $w : D \to B(0, R)$ locally uniform on D.

Fix subdomains U and D_1 of the domain D such that $U \subset\subset D_1 \subset\subset D$. We have

$$\sup_{x \in D_1} |w_n(x)| \leq C(D_1) \quad \text{for all} \quad n = 1, 2, \ldots$$

with a constant $C(D_1) < \infty$. Thus from (3.20) it follows

$$\iint\limits_{U} \Lambda(x, w_n) \, dx_1 dx_2 \leq C_1(U, D) \iint\limits_{D_1} \left(K^2 + |\nabla K|^2 \right) \, dx_1 dx_2$$

where $C_1(U, D) = c_1 \, C^2(D_1)$ is a constant.

In other words, the sequence of $W^{1,2}$-homeomorphisms w_n has uniform bounded Dirichlét integrals on the subdomain U. Because the class of maps with uniform bounded Dirichlét integrals is closed with respect to locally uniform convergence (see, for example, Theorem 1 of the monograph of Suvorov [177, Chapter I, §1]), then we are able to conclude that the limiting map $w : D \to B(0, R)$ belongs to $W^{1,2}_{\text{loc}}(D)$.

Suppose that the vector function $f^{-1} : \Omega \to D$ belongs to $W^{1,\alpha}_{\text{loc}}(\Omega)$ for some

$2 \leq \alpha < \infty$. Then for an arbitrary subdomain $\Omega_1 \subset\subset \Omega$, we have

$$\iint\limits_{\Omega_1} \Lambda^{1/2}(y, F^{-1})\, d\operatorname{area}_\Omega \leq \iint\limits_{\Omega_1} \Lambda^{1/2}(x, w)\, \Lambda^{1/2}(y, f^{-1})\, d\operatorname{area}_\Omega =$$

$$= \iint\limits_{U} \Lambda^{1/2}(x, w)\, \Lambda^{1/2}(y, f^{-1}) \sqrt{g_{11}g_{22} - g_{12}^2}\, dx_1 dx_2 \leq$$

$$\leq \left(\iint\limits_{U} \Lambda(x, w)\, dx_1 dx_2 \right)^{1/2} \times$$

$$\times \left(\iint\limits_{U} \Lambda^{\alpha/2}(y, f^{-1}) \sqrt{g_{11}g_{22} - g_{12}^2}\, dx_1 dx_2 \right)^{1/\alpha} \times$$

$$\times \left(\iint\limits_{U} \left(g_{11}g_{22} - g_{12}^2 \right)^{\beta/2} dx_1 dx_2 \right)^{1/\beta}.$$

Here $U = f^{-1}(\Omega_1)$ and $U \Subset D$. Thus the assumption (3.13) implies that F^{-1} belongs to $W_{\text{loc}}^{1,1}(\Omega)$. It proves the statement (iii) of Theorem.

Now as above, we observe that the sequence of maps $F_n : B_n \to \Omega$, $n = 1, 2, \ldots$ converges to the homeomorphism $F : B(0, R) \to \Omega$ uniformly on every subdomain $\Delta' \Subset B(0, R)$. Since $F_n = f \circ w_n^{-1}$, then the sequence of Q_n-quasiconformal maps $w_n^{-1} = f^{-1} \circ F_n$ converges locally uniformly on the kernel $B(0, R)$ to the homeomorphism $w^{-1} : B(0, R) \to D$. Using the estimate (3.26) and the closure of maps with uniform bounded Dirichlét with respect to uniform convergence, as in the previous case, we conclude that $w^{-1} \in W_{\text{loc}}^{1,2}(B(0, R))$.

Thus both mappings w and w^{-1} belong to $W_{\text{loc}}^{1,2}$ on the corresponding domain and, consequently, have Lusin N-property (see, for example, Maly [97, Theorem B]). This implies the property (i) of Theorem.

Now we have to prove the property (ii). Putting $w(U) = \Delta'$, we have

$$\iint\limits_{\Delta'} \Lambda^{1/2}(\xi, F)\, d\xi_1 d\xi_2 \leq \iint\limits_{\Delta'} \Lambda^{1/2}(x, f)\, \Lambda^{1/2}(\xi, w^{-1})\, d\xi_1 d\xi_2 \leq$$

$$\leq \left(\iint\limits_{\Delta'} \Lambda(x, f)\, J(\xi, w^{-1})\, d\xi_1 d\xi_2 \right)^{1/2} \left(\iint\limits_{\Delta'} \frac{\Lambda(\xi, w^{-1})}{J(\xi, w^{-1})}\, d\xi_1 d\xi_2 \right)^{1/2} \leq$$

$$\leq \left(\iint\limits_{U} \Lambda(x, f)\, dx_1 dx_2 \right)^{1/2} \left(\iint\limits_{\Delta'} \frac{\Lambda(\xi, w^{-1})}{J(\xi, w^{-1})}\, d\xi_1 d\xi_2 \right)^{1/2}.$$

Here we used also the change variables formula (3.27) what is possible since $w^{-1} \in W^{1,2}_{\mathrm{loc}}(B(0, R))$ (see, for example, Heinonen, Koskela [60, Theorem **6.1**]).
   Now observe that

$$\frac{\Lambda(\xi, w^{-1})}{J(\xi, w^{-1})} \leq 2p(\xi, w^{-1})$$

and for $\xi = w(x)$ it is fulfilled $p(\xi, w^{-1}) = p(x)$. From this by Lemma 3.19, we conclude that

$$\iint\limits_{\Delta'} \frac{\Lambda(\xi, w^{-1})}{J(\xi, w^{-1})}\, d\xi_1 d\xi_2 \leq 2 \iint\limits_{U} p(x)\, J(x, w)\, dx_1 dx_2 \leq$$

$$\leq 2 \iint\limits_{U} \Lambda(x, w)\, dx_1 dx_2 < \infty.$$

Thus $F \in W^{1,1}_{\mathrm{loc}}(B(0, R))$ and Theorem is proved completely.   □

3.5   Proof of Theorem 3.8

Let $\xi_0 \in \mathbb{R}^2$ be a point and let $0 < t' < t'' < \infty$ be a pair of fixed numbers. Denote by

$$S(\xi_0, r) = \{\xi \in \mathbb{R}^2 : |\xi - \xi_0| = r\}$$

the circle with the center ξ_0 and the radius $t' < r < t''$.

Let V be a domain in $\mathbb{R}^2$. For an arbitrary r, $t' < r < t''$, let $S_V(\xi_0, r)$ be a connected component of the set $S(\xi_0, r) \cap V$. Moreover, we assume that these components are chosen such that the set

$$K_V(\xi_0, t', t'') = \cup_{t' < r < t''} S_V(\xi_0, r)$$

is connected subset of the domain V.

We will need a corollary from the Length and Area Principle (see Theorem 1.51 or [177, Part I, Chapter III, §6]).

Lemma 3.29 *Let $g : V \to \mathbb{R}^2$ be a vector function of the class $W^{1,2}(V)$. The following estimate holds*

$$\int\limits_{t'}^{t''} \mathrm{osc}^2\left(g, S_V(\xi_0, r)\right) \frac{dr}{r} \leq c_3\, I(g) \qquad (3.30)$$

where c_3 is an absolute constant and

$$I(g) = \int\limits_{K_V} \Lambda(\xi, w_n^{-1})\, d\xi_1 d\xi_2, \quad K_V = K_V(\xi_0, t', t'').$$

Let $\{F_n : B_n \to \Omega\}$ be a sequence of canonical representations of the surface Ω corresponding to the sequence $\{Q_n\}$, $n = 1, 2, \ldots$. Let $F_n = f \circ w_n^{-1}$ where $w_n : D \to B_n$ are Q_n-quasi-conformal mappings normed by conditions (3.6). Here $B_n = B(0, R_n)$ and $1 < R_n \leq \infty$.

At first we show

$$\overline{\lim_{n \to \infty}} R_n = R < \infty. \qquad (3.31)$$

We assume that (3.31) is not true. Choose in Lemma 3.29 the point $\xi_0 = 0$, numbers $t' = 1$, $t'' = R_n > 1$ and $g = w_n^{-1}$. The inequality (3.30) implies

$$\inf_{1 < r < R_n} \mathrm{osc}\,(w_n^{-1}, S(0, r)) \leq c_3^{1/2}\, I^{1/2}(w_n^{-1}) \ln^{-1/2} R_n.$$

The mappings w_n^{-1} are homeomorphic, and consequently, for every $r \in (t', t'')$ we have

$$\mathrm{osc}\,\left(w_n^{-1}, B(0, r)\right) \leq \mathrm{osc}\,\left(w_n^{-1}, S(0, r)\right).$$

Choosing in Lemma 3.25 subsets $U = D$ and $V = B_n$, we obtain

$$I(w_n^{-1}) \leq \int\limits_{D} K(x)\, dx_1 dx_2 \equiv A(D). \qquad (3.32)$$

By (3.6) for every $r > 1$, points a_0, a_1 lie in domains $w_n^{-1}(B(0,r))$, and hence,

$$|a_1 - a_0| \leq \operatorname{osc}\left(w_n^{-1}, B(0,r)\right), \quad 1 < r < R_n.$$

Combining (3.30) and (3.32), we arrive to

$$|a_1 - a_0| \leq c_3^{1/2} A^{1/2}(D) \ln^{-1/2} R_n. \tag{3.33}$$

For sufficiently large $R_n > 1$ the inequality (3.33) is not valid. Thus the relation (3.31) is proved. Moreover, the inequality (3.33) can be a source of nontrivial upper estimates for radii R_n. Namely, from (3.33) it follows that

$$\ln R_n \leq \frac{c_3 A(D)}{|a_1 - a_0|^2} \quad (n = 1, 2, \ldots).$$

Setting $n \to \infty$, we have

$$\ln R \leq \frac{c_3 A(D)}{|a_1 - a_0|^2}. \tag{3.34}$$

Now we prove that

$$\varliminf_{n \to \infty} R_n > 1. \tag{3.35}$$

We use Lemma 3.29. Fix $n = 1, 2, \ldots$. Choose $V = B(0, R_n)$ and $g = w_n^{-1}$. Consider the family of circles with the centers at $a_n = (R_n, 0)$ and radii $t' < r < t''$ where $t' = R_n - 1$, $t'' = R_n$. By (3.30) we can write

$$\inf_{t' < r < t''} \operatorname{osc}\left(w_n^{-1}, S_V(a_n, r)\right) \leq c_3^{1/2} I^{1/2}(w_n^{-1}) \ln^{-1/2} \frac{R_n}{R_n - 1}.$$

Using (3.32), we conclude on existence

$$\bar{r}, \quad R_n - 1 < \bar{r} < R_n$$

such that

$$\operatorname{osc}\left(w_n^{-1}, S_V(a_n, \bar{r})\right) \leq c_3^{1/2} A^{1/2}(D) \ln^{-1/2} \frac{R_n}{R_n - 1}.$$

Thus on the domain D there is an arc

$$\gamma = w_n^{-1}\left(S_V(a_n, \bar{r})\right)$$

with the diameter $d(\gamma)$ satisfying the inequality

$$d(\gamma) \leq c_3^{1/2} A^{1/2}(D) \ln^{-1/2} \frac{R_n}{R_n - 1}. \tag{3.36}$$

The arc γ separates on D the point a_1 from the point a_0 and has endpoints on ∂D. Let $\delta(a_0, a_1)$ be the infimum of the diameters of these arcs γ. Thus from (3.36) it follows that

$$\ln \frac{R_n}{R_n - 1} \leq \frac{c_3 A(D)}{\delta^2(a_0, a_1)} \quad (n = 1, 2, \ldots).$$

Letting $n \to \infty$, we have

$$\frac{\lambda}{\lambda - 1} \leq R \quad \text{where} \quad \lambda = \exp\left\{\frac{c_3 A(D)}{\delta^2(a_0, a_1)}\right\}. \tag{3.37}$$

This proves (3.35).

Now we prove that the family of maps w_n^{-1} is equipotential uniform continuous inside B_n, $n = 1, 2, \ldots$. Let $V \subset\subset B_n$ be a subdomain of the disc B_n, containing points $\xi = (0, 0)$ and $\xi = (1, 0)$. Choose a pair of points ξ', $\xi'' \in V$ so that

$$|\xi'' - \xi'| < \min\{1, \text{dist}^2(V, \partial B_n)\}. \tag{3.38}$$

We consider a family of circles $\{S(\xi', r)\}$ with the center at ξ' and the radius $r \in (t', t'')$ where

$$t' = |\xi'' - \xi'|, \quad t'' = |\xi'' - \xi'|^{1/2}.$$

The estimate (3.30), applied to the vector function w_n^{-1} and the domain $B_n = B(0, R_n)$, implies

$$\inf_{t' < r < t''} \text{osc}\,(w_n^{-1}, S(\xi', r)) \leq c_3^{1/2} I^{1/2}(w_n^{-1}) \ln^{-1/2} \frac{1}{|\xi'' - \xi'|^{1/2}}.$$

As above, using (3.32), we conclude validity of the following statement.

Lemma 3.39 *If a characteristic $\mathcal{P}(x)$ is $W^{1,2}$-majorized on a domain $D \subset \mathbb{R}^2$, then for every pair of points ξ', $\xi'' \in V$ with (3.38), it is fulfilled*

$$|w_n^{-1}(\xi'') - w_n^{-1}(\xi')| \leq c_3^{1/2} A^{1/2}(D) \ln^{-1/2} \frac{1}{|\xi'' - \xi'|^{1/2}}. \tag{3.40}$$

Now we are going to prove the equipotential uniform continuity of the mappings w_n $(n = 1, 2, \ldots)$ on compact subsets of D. Fix subdomains $U \Subset D$ and D_1 such that

$$U \Subset D_1 \Subset D.$$

By (3.31) there exists a number N such that for all $n > N$ it is fulfilled $R_n \leq 2R$. By Lemma 3.19 for every $n > N$, we have

$$\iint_{D_1} \Lambda(x, w_n)\, dx_1 dx_2 \leq 2\, c_1\, R \iint_D (K^2 + |\nabla K|^2)\, dx_1 dx_2 \equiv B(D_1, D)\, \|K\|$$

where $B(D_1, D) \|K\|$ is a constant depending only on $\|K\| \equiv \|K\|_{W^{1,2}(D)}$ and subdomains D_1, D.

Let x', $x'' \in U$ be a pair of points for which

$$|x'' - x'| \leq \min\{1, \mathrm{dist}^2(U, \partial D_1)\}. \tag{3.41}$$

As above, we prove the following lemma.

Lemma 3.42   *Assume that a function $\mathcal{P}(x)$ is $W^{1,2}$-majorized on a domain $D \subset \mathbb{R}^2$. If points x', $x'' \in U$ satisfy (3.41), then for every $n > N$, it is fulfilled*

$$|w_n(x'') - w_n(x')| \leq c_3^{1/2}\, B^{1/2}(D_1, D)\, \ln^{-1/2} \frac{1}{|x'' - x'|}. \tag{3.43}$$

We choose a subsequence $\{R_{n_k}\}$ with a limit R, $1 < R < \infty$. The corresponding sequence of discs $\{B(0, R_{n_k})\}$ converges to the kernel $B(0, R)$ with respect to the point $\xi = (0,0)$ (see, for example, [177, Chapter II, §3] for the definition of the kernel convergence and properties of the kernel). The subsequence of maps $F_{n_k} = f \circ w_{n_k}^{-1} : B(0, R_{n_k}) \to D$ is the canonical sequence of representations of Ω corresponding to the sequence of numbers Q_{n_k}.

The estimate (3.43) means that the sequence $\{w_{n_k}\}$ is equipotential uniform continuous on every compact subset of D. By the standard diagonal process, we can choose a subsequence $\{w_{n'_k}\}$ which converges to a continuous map $w : D \to \mathbb{R}^2$ locally uniform.

The inequality (3.40) of Lemma 3.39 states that the family $\{w_{n'_k}^{-1}\}$ of inverse maps is uniformly continuous on every compact subset of $B(0, R)$, and hence, the limiting map w for $w_{n'_k}$ is homeomorphic. It is clear that w satisfies (3.6) also. Moreover, by closeness of the class $W^{1,2}$-functions with a bounded Dirichlét integral, using (3.20) and (3.26), we conclude that $w \in W^{1,2}_{\mathrm{loc}}(D)$, and $w^{-1} \in W^{1,2}(B(0, R))$.

Lemma 3.44   *Let $g_k(x) : D \to \mathbf{R}^2$, $k = 1, 2, \ldots$, be a sequence of Q_k-quasi-conformal homeomorphisms with characteristics $(p_k(x), \theta_k(x))$ which converges locally uniform in D to a $W^{1,2}_{\mathrm{loc}}(D)$-homeomorphism with characteristics $(p(x), \theta(x))$.*

If for every subdomain $U \subset\subset D$ there exists a constant $M(U)$, $0 < M(U) < \infty$ such that

$$\int\limits_U \Lambda(x, g_k)\, dx_1 dx_2 \leq M(U) \quad (k = 1, 2, \ldots) \tag{3.45}$$

and a.e. in D:

$$(p_k(x), \theta_k(x)) \to (p_\infty(x), \theta_\infty(x)), \quad k \to \infty,$$

then $(p_\infty(x), \theta_\infty(x)) = (p(x), \theta(x))$ a.e. in D.

This statement is a special case of Theorem 5.21 which will be proved in Section 5.3.

The sequence $w_{n'_k}(x)$ of $W^{1,2}$-homeomorphisms converges as $n'_k \to \infty$ to a homeomorphism $w(x) : D \to B(0, R)$ locally uniform on D. By Lemma 3.26, the property (3.45) is valid. Other assumptions of Lemma 3.44 are valid by virtue of the construction. Thus the limiting homeomorphism $w(x) : D \to B(0, R)$ has characteristics $(p(x), \theta(x))$ a.e. on D.

It is not difficult to see that the vector function $F = f \circ w^{-1}$ defines isothermal coordinates on Ω. Indeed, the vector function $f : D \to \mathbb{R}^m$ is differentiable a.e. on D and satisfies (3.1). The maps w, w^{-1} are differentiable a.e. on domains D and $B(0, R)$ respectively, and they also have Lusin N-property. Thus, obviously, the composition $F = f \circ w^{-1}$ is differentiable a.e. and satisfies to (3.1) a.e. on $B(0, R)$. Moreover, at almost every point where it has a differential, it transforms infinitesimal circles on $B(0, R)$ into infinitesimal ellipses on Ω. This means that a.e. on D the vector function F satisfies (2.3).

We will prove the uniqueness of the found isothermal representation of Ω. Suppose that there are two different such canonical homeomorphisms F_1 and F_2. This means that there exist two different quasi-conformal mappings $g : D \to B(0, R')$ and $h : D \to B(0, R'')$ satisfying to (3.6) and having one pair of characteristics (p, θ). We can write

$$F_1 = f \circ g^{-1} : B(0, R') \to \Omega, \quad F_2 = f \circ h^{-1} : B(0, R'') \to \Omega.$$

It is necessary to prove that the mapping $\varphi = F_1^{-1} \circ F_2$ is identical. We have

$$\varphi = (g \circ f^{-1}) \circ (f \circ h^{-1}) = g \circ h^{-1}.$$

Since g and h^{-1} belong to $W_{loc}^{1,2}$, then $\varphi \in W_{loc}^{1,1}(B(0, R''))$. On the other hand, a.e. the map φ transforms infinitesimal circles into infinitesimal circles. Thus this map is a qr-mapping and, hence, is conformal in traditional sense (see Koskela and Malý [71, Teorem **1.1**] or Iwaniec and Martin [64, Corollary **5.3.1**] and the definition of qr-maps with minimal assumptions in the same place). The conformal transformation φ maps the disc $B(0, R'')$ onto the disc $B(0, R')$ under fixed points $(0, 0)$ and $(1, 0)$. It is possible if and only if $R' = R''$ and $\varphi \equiv \xi$. That is the map $\varphi = F_1^{-1} \circ F_2$ is identical. Thus $F_1 \equiv F_2$.

The uniqueness of the canonical homeomorphism

$$F = f \circ w^{-1} : B(0, R) \to \Omega$$

implies in turn that every subsequence

$$\{F_{n_k} : B(0, R_{n_k}) \to D\}$$

of the sequence of homeomorphisms

$$\{F_n = f \circ w_n^{-1} : B(0, R_n) \to D\}$$

has the same map F. It means, in particular, that

$$\lim_{n \to \infty} R_n = \overline{\lim_{n \to \infty}} R_n$$

and that the sequence of domains B_n converges to the kernel $B = B(0, R)$. Theorem is proved completely. $\square$

3.6 Bi-Lipschitz Surfaces

Let $\Omega \subset \mathbb{R}^m$ be a surface given by a vector function (2.1). Suppose that the surface is locally bi-Lipschitz. For a point $x \in D$, we put

$$\lambda(x) = \lim_{x' \to x} \frac{|f(x') - f(x)|}{|x' - x|} \quad \text{and} \quad \Lambda(x) = \overline{\lim_{x' \to x}} \frac{|f(x') - f(x)|}{|x' - x|}.$$

The assumption of locally bi-Lipschitz implies

$$0 < \lambda(x) \le \Lambda(x) < \infty \quad (x \in D).$$

On the other hand, for coefficients g_{11} and g_{22} of the first quadratic form of Ω at every point $x \in D$ where f is differentiable, we have

$$g_{11} = \left| \frac{\partial f}{\partial x_1} \right|^2 \le \Lambda^2, \quad g_{22} = \left| \frac{\partial f}{\partial x_2} \right|^2 \le \Lambda^2.$$

It is not difficult to estimate the coefficient of the area distortion for $f : D \to \Omega$. Indeed, for every point $x \in D$ at which f is differentiable, and for each disc $B(x, r) \subset D$, we have

$$\text{mes}_2 \, f(B(x, r)) = \iint\limits_{B(x,r)} \sqrt{g_{11} g_{22} - g_{12}^2} \, dx_1 dx_2 \ge \pi \min_{|x' - x| = r} |f(x') - f(x)|^2.$$

From this we find

$$\lim_{r \to 0} \frac{1}{r^2} \iint\limits_{B(x,r)} \sqrt{g_{11} g_{22} - g_{12}^2} \, dx_1 dx_2 \ge \pi \lim_{r \to 0} \frac{|f(x') - f(x)|^2}{r^2},$$

and a.e. on D, it is fulfilled

$$\sqrt{g_{11} g_{22} - g_{12}^2} \ge \lambda^2.$$

Thus for the function $\mathcal{P}(x)$ defined by (3.9), we have

$$\mathcal{P}(x) \leq \frac{2\Lambda^2(x)}{\lambda^2(x)}. \qquad (3.46)$$

In the considered case, we can combine Theorems 3.8 and 3.12 by the following way.

Theorem 3.47 *Let Ω be a two-dimensional simply connected surface, embedded in $\mathbb{R}^m$ by a locally bi-Lipschitz vector function (2.1) given on a domain $D \subset \mathbb{R}^2$. Assume that the function $\Lambda^2(x)/\lambda^2(x)$ is $W^{1,2}$-majorized on D.*
   Then there exist isothermal coordinates $\xi = (\xi_1, \xi_2)$ on Ω defined by a canonical homeomorphism $y = F(\xi) : B(0, R) \to \mathbb{R}^m$ and having the properties described in Theorem 3.8.
   Moreover, $F \in W^{1,2}_{\mathrm{loc}}(B(0,R))$, and $F^{-1} \in W^{1,2}_{\mathrm{loc}}(\Omega)$.

Proof. It is sufficient to observe that (3.46) implies the $W^{1,2}$-majorability on D the function (3.9) and, thus, the applicability of Theorem 3.8.
   Because $F = f \circ w^{-1}$ where w and w^{-1} are $W^{1,2}_{\mathrm{loc}}$-homeomorphisms and f is the locally bi-Lipschitz map, then obviously, F has the necessary properties. $\square$

3.7   Quasi-conformal Mappings

Let $D \subset \mathbb{R}^2$ be a simply connected domain and let $(p(x), \theta(x))$ be a pair of measurable characteristics where $p(x)$ is $W^{1,2}$-majorized on D. As in the proof of Theorem 3.8, we verify that the following statement is valid.

Theorem 3.48 *There exists a homeomorphic mapping $\xi = w(x) : D \to \mathbb{R}^2$ with properties:*
   (i) w, $w^{-1} \in W^{1,2}$ on D and $\mathcal{D} = w(D)$ respectively;
   (ii) a.e. on D the mapping $w(x)$ has characteristics $(p(x), \theta(x))$.

A corresponding result is valid also in the case of domains with any connectedness [98].

Theorem 3.49 *Let D be a subdomain of $\mathbb{R}^2$ and let $(p(x), \theta(x))$ be measurable characteristics on D where $p(x)$ is locally $W^{1,2}$-majorized on D.*
   Then there exists a homeomorphism $\xi = w(x) : D \to \mathbb{R}^2$ with properties:
   (i) w, $w^{-1} \in W^{1,2}_{\mathrm{loc}}$ on domains D and $\mathcal{D} = w(D)$ respectively;
   (ii) a.e. on D the mapping $w(x)$ has characteristics $(p(x), \theta(x))$;
   (iii) the mapping $\xi = w(x)$ is defined up to conformal mappings on the (ξ_1, ξ_2)-plane.

Proof. Let $\{D_n\}_{n=1}^\infty$ be a sequence of subdomains D such that

$$D_1 \Subset D_2 \ldots \Subset D_n \ldots, \quad \cup_{n=1}^\infty D_n = D.$$

Let $R_n = \max_{x \in \overline{D}_n} |x|$ and let $p_n(x) = p(x)$ for $x \in D_n$, $p_n(x) = 1$ for $x \in \mathbb{R}^2 \setminus D_n$. Characteristics $(p_n(x), \theta(x))$ satisfy the assumptions of Theorem 3.48 on every disc $|x| < R$. Fix $a_1 \in D_1$, $a_1 \neq 0$. By Theorem 3.48 there exists a homeomorphism $\xi = w_n(x)$, $w_n(0) = 0$, $w_n(a_1) = a_1$ which maps the disc $|x| < R_n$ onto a suitable disc $|\xi| < r_n$ with characteristics $(p_n(x), \theta(x))$ a.e. and such that $w_n \in W_{\mathrm{loc}}^{1,2}(B(R_n))$, $w_n^{-1} \in W_{\mathrm{loc}}^{1,2}(B(r_n))$.

On every subdomain D_{n_0}, the maps $w_n(x)$, $n > n_0$ can be represented in the form $w_n = \varphi_n \circ w_{n_0}(x)$ where φ_n are conformal on the domain $\tilde{D}_{n_0} = w_{n_0}(D_{n_0})$. Since $\varphi_n(0) = 0$ and $\varphi_n(a_1) = a_1$, then Theorem 4.1 from [90] guarantees the existence of a subsequence $\{\varphi_{n'}\}$ locally uniform converging on $\tilde{D}_{n_0}$ to a conformal map $\varphi : \tilde{D}_{n_0} \to \mathbb{R}^2$. Thus $\{w_n\}$ contains a subsequence $\{w_{n'}\}$ of quasi-conformal maps which converges locally uniform on $\tilde{D}_{n_0}$ to a homeomorphism $w_0 = \varphi \circ w_{n_0}$. By Lemma 3.44 the map w_0 has characteristics $(p(x), \theta(x))$ a.e. on $\tilde{D}_{n_0}$.

Using the diagonal process, we find a subsequence $w_{n''}$ locally uniform, converging on D to a homeomorphism $w : D \to \mathbb{R}^2$ with characteristics $(p(x), \theta(x))$. It is clear that $w \in W_{\mathrm{loc}}^{1,2}(D)$ and $w^{-1} \in W_{\mathrm{loc}}^{1,2}(\mathcal{D})$. This map is unique up to conformal transforms on the ξ-plane. $\square$

Open questions 3.50   1) What are isothermal coordinates in the case of Finsler's metrics ? 2) Find conditions under which there exist isothermal coordinates on abstract surfaces of the general form. 3) Prove main results of this chapter under smaller restrictions for Ω.

Chapter 4

The Boundary of a Surface

Below, we introduce simple ends on two-dimensional simply connected surfaces in $\mathbb{R}^m$ which are analogous to Carathéodory prime ends of plane domains (see, for example, [177, Chapter I, §3]). Estimates of the Lavrentiev relative distance are given under conformal mappings of locally bi-Lipschitz surfaces [119].

4.1 The Relative Distance

Let $D \subset \mathbb{R}^2$ be a simply connected domain and $O \in D$ be a fixed point. Let $U \subset D$ be an open set.

We define some notations. By the symbol $\overline{U}$, we denote a closure of U with respect to the topology of $\mathbb{R}^2$. Now we put $[U] = \overline{U} \setminus \partial D$ and $\partial' U = [U] \setminus U$.

We consider a locally Lipschitz surface Ω embedded into $\mathbb{R}^m$ with a vector function (3.14).

Let $D^* = f(D) \subset \Omega$ be a simply connected domain and $O^* = f(O)$. If points $a, b \in D'$, $a, b \neq O^*$, then let

$$\rho(a, b; O^*, D^*) = \min\{\rho_1(a, b),\ \rho_2(a, b)\} \tag{4.1}$$

where ρ_1 is the infimum of lengths (with respect to the metric of $\mathbb{R}^m$) of the closed curves $\gamma \subset D^* \setminus \{O^*\}$ separating[1] a and b from the point O^* and the boundary ∂D^*; ρ_2 is the infimum of lengths of the arcs lying on $D^* \setminus \{O^*\}$ and separating a and b from O^* on the surface D^*.

The quantity ρ is called the *relative distance* between points $a, b \in D^* \setminus \{O^*\}$.

---

[1] An arc (or curve) $\gamma \subset D$ separates a point $a \in D$ from a point $b \in D$ if for every path $l \subset D$ joining on D points a and b it is fulfilled $l \cap \gamma \neq \emptyset$.

A relative distance from $a \in D^* \setminus \{O^*\}$ to O^* is defined by the relation

$$\rho(a, O^*; O^*, D^*) = \lim_{\substack{b \to O^* \\ b \neq O^*}} \rho(a, b; O^*, D^*).$$

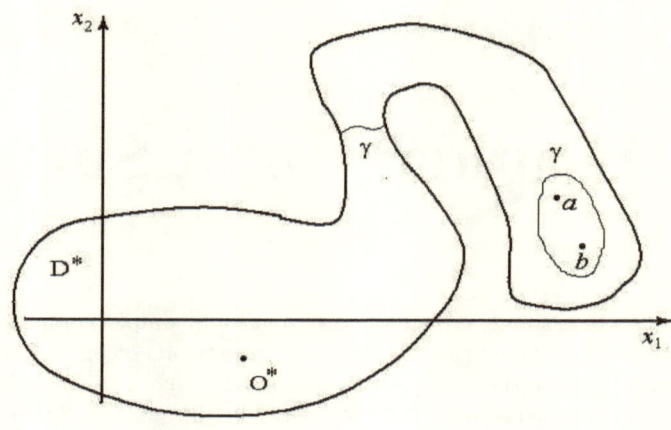

Fig 4.1.

For a plane two-dimensional surface $\Omega \subset \mathbb{R}^m$, obviously, we have

$$\rho(a, b; O^*, D^*) = \min\{2\rho_1^*(a, b), \rho_2(a, b)\} \qquad (4.2)$$

where ρ_1^* is the infimum of lengths of the arcs $\gamma \subset D^* \setminus \{O^*\}$ joining points a and b on D^*.

The relative distance (4.2) (with no factor 2 before ρ_1^*) had been introduced by Lavrentiev [85]. Suvorov [176] constructed a counterexample, showing that Lavrentiev distance does not satisfy the triangle axiom. He replaced lengths with diameters. However, it is possible to correct this mistake by smaller radical means. Namely, the following result holds.

Theorem 4.3 *If Ω is a surface given by a vector function (3.14), then the function (4.1) defines a metric on $D \subset \Omega$ which satisfies the axioms of symmetry, identity and triangle.*

Proof. Since the surface Ω is locally Lipschitz, then for every pair of points $a, b \in D$ it is fulfilled $\rho(a, b; O, D) < \infty$. The symmetry axiom is trivial. We prove that $\rho(a, b; O, D) = 0$ if and only if $a = b$.

Indeed, we assume that $a \neq b$. Let $x' = y^{-1}(a)$, $x'' = y^{-1}(b)$ be preimages of these points on a simply connected domain $\Delta^* = y^{-1}(D)$ and let $O^* = y^{-1}(O)$. The map $y : \Delta^* \to D$ is homeomorphic, and hence, $x' \neq x''$. For plane domains the identity axiom is valid. Thus $\rho(x', x''; O^*, \Delta^*) > 0$.

Assumptions on the homeomorphism $y = y(x)$ and simply connectedness of D imply that $\rho(a, b; O, D) > 0$.

Now it is sufficient to check the triangle inequality. Fix points $a, b, c \in \Omega \setminus \{O\}$. There are three cases.

In the *first case*, we have

$$\rho(a, b; O, D) = \rho_1(a, b), \quad \rho(b, c; O, D) = \rho_1(b, c).$$

Fix $\varepsilon > 0$. Choose closed curves γ_1 separating a, b from O, ∂D and γ_2 separating b, c from O, ∂D such that

$$\text{length}(\gamma_1) \leq \rho_1(a, b) + \frac{\varepsilon}{2}, \quad \text{length}(\gamma_2) \leq \rho_1(b, c) + \frac{\varepsilon}{2}.$$

If $\gamma_1 \cap \gamma_2 = \emptyset$, then at least one among curves separates other curve. It means that this curve separates points a, c from O and ∂D. Thus for an arbitrary $\varepsilon > 0$ we have

$$\rho_1(a, c) \quad \leq \max\{\text{length}(\gamma_1), \text{length}(\gamma_2)\} \leq$$

$$\leq \rho_1(a, b) + \rho_1(b, c) + \varepsilon \leq$$

$$\leq \rho(a, b; O, D) + \rho(b, c; O, D) + \varepsilon.$$

If $\gamma_1 \cap \gamma_2 \neq \emptyset$, then the set $\gamma_3 = \gamma_1 \cup \gamma_2$ would be a connected closed curve separating a, c from O and ∂D. Therefore,

$$\rho_1(a, c) \leq \quad \rho_1(a, b) + \rho_1(b, c) + \varepsilon \leq$$

$$\leq \rho(a, b; O, D) + \rho(b, c; O, D) + \varepsilon.$$

Setting $\varepsilon \to 0$, we arrive to the necessary inequality.

In the *second case*, let

$$\rho(a, b; O, D) = \rho_2(a, b), \quad \rho(b, c; O, D) = \rho_2(b, c).$$

Choose open arcs γ_1, γ_2, separating a, b and b, c from O, respectively on D. Moreover, for some $\varepsilon > 0$ it is fulfilled

$$\text{length}(\gamma_1) \leq \rho_2(a, b) + \frac{\varepsilon}{2}, \quad \text{length}(\gamma_2) \leq \rho_2(b, c) + \frac{\varepsilon}{2}.$$

Here as above, there are two subcases. In both variants we obtain necessary inequalities.

In the *third case,* it is sufficient to consider the case in which

$$\rho(a,b;O,D) = \rho_1(a,b), \quad \rho(b,c;O,D) = \rho_2(b,c),$$

and to choose

$$\text{length}(\gamma_1) \le \rho_1(a,b) + \frac{\varepsilon}{2}, \quad \text{length}(\gamma_2) \le \rho_2(b,c) + \frac{\varepsilon}{2}.$$

Further arguments are as above.   □

4.2 Prime Ends

We use the well-known scheme [177, Part I, §3] of the introduction of Carathéodory prime ends on plane simply connected domains [24]. Let Ω be a surface embedded into $\mathbb{R}^m$ with a vector function (3.14). Consider a simply connected subdomain $D \subset \Omega$ with a fixed point $O \in D$.

Let ρ be a relative distance on D defined by (4.1). The metric space (D, ρ) can be completed with equivalence classes of the fundamental sequences $\{a_k\}$, $a_k \in D$ which do not have limiting points on D. The equivalence classes $e^\rho = \{a_k\}$ of such sequences are called the *prime ends* of the domain D (with respect to the metric ρ). By $\tilde{D}^\rho$ we denote the domain D with the joined prime ends e^ρ.

Let $e_1^\rho = \{a_k'\}$, $e_2^\rho = \{a_k''\}$ be points of $\tilde{D}^\rho$. The relative distance between them is defined by the following formula

$$\rho(e_1^\rho, e_2^\rho; O, \tilde{D}^\rho) = \lim_{k \to \infty} \rho(a_k', a_k''; O, D).$$

If we fix other inner points of D instead of $O \in D$, then we obtain different metrics ρ on D and different metric spaces D^ρ. The supplements $\tilde{D}^\rho \setminus D$ of these spaces are identical.

Indeed, it is sufficient to observe that every sequence $\{a_k\}$ of the surface D which has no limiting points on D and is fundamental with respect to $\rho(a,b;O',D)$, is also fundamental with respect to $\rho(a,b;O'',D)$.

If D is a simply connected proper subdomain $\mathbb{R}^2$, then the metric ρ is defined by (4.2). It is not difficult to check that, in this case, the simple ends e^ρ of D are simple ends of Carathéodory [24], [176].

Next we will follow the original article of Luferenko and Suvorov [94] where this concept is studied in arbitrary metric spaces. For other definitions of conformally invariant boundary elements, see Ivanov and Suvorov [65], Karmazin [67].

Let $e_0 \in \tilde{D}$ be a simple end and $y_0 \in \mathbb{R}^m$ be a point. We write $y_0|e_0 \neq \emptyset$ if there exists a sequence of points $a_k \in D$, $\rho(a_k, e_0; O, D) \to 0$ with $|a_k - y_0| \to 0$. The set of all points $y \in \mathbb{R}^m$ such that $y|e_0 \neq \emptyset$ is called the *body* $|e_0|$ of a simple end $e_0 \in \tilde{D}$ in $\mathbb{R}^m$.

This definition is the extension of the Carathéodory concept [24].

It is not difficult to prove that

$$|e_0| = \cap_{\delta>0} \overline{\{y \in D : \rho(y, e_0) < \delta\}}.$$

(Here the closings are with respect to $\mathbb{R}^m$.)

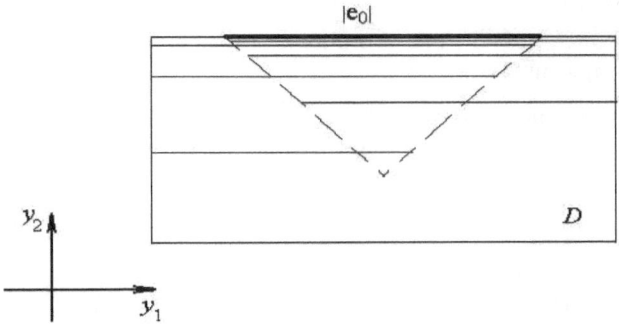

Fig. 4.2.

For the case of metric spaces, see the proof [94, Lemma 1].

The classification of the prime ends of a simply connected surface D is the same as in the Carathéodory theory. Namely, we call by a *cross section of a surface* D a Jordan arc $\gamma \subset D$ with ends on the Euclidean boundary ∂D. A cross-section γ *divides* points $a, b \in D$ if every path $l \subset D$ conducting from a to b intersects γ. A cross-section $\gamma \subset D$ *separates* a prime end $e \in \tilde{D} \setminus D$ from a fixed point $O \in D$ if it divides the point O and other points on the surface D which are sufficiently close to e with respect to the relative distance.

We will say, that a *sequence of cross-sections* $\gamma_k \subset D$ separating a prime end e from a fixed point $O \in D$ *shrinks* to e if

$$r(\gamma_k, e; D) = \inf_{y \in \gamma_k} \overline{\lim}_{y_l \to e} |y - y_l| \to 0$$

as $k \to \infty$.

A *principal point* of a body of a prime end $e \in \tilde{D} \setminus D$ is the point $y \in |e|$ for which a sequence of cross-sections γ_k shrinks to e and has the property

$$\sup_{a \in \gamma_k} |a - y| \to 0 \quad \text{as} \quad k \to \infty.$$

An *adjacent point* of a body of $e \in \tilde{D} \setminus D$ is its point which is not principal.

Every prime end of a simply connected surface belongs to one of the following four types:

a *prime end of the I-st type* contains a unique principal point and does not contain adjacent points;

a *prime end of the II-nd type* contains a unique principal point and more than one adjacent point;

a *prime end of the III-rd type* contains more than one principal point and does not contain adjacent points;

a *prime end of the IV-th type* contains more than one principal point and more than one adjacent point.

It is easy to construct examples of surfaces with prime ends of the indicated above types. In particular, they are admissible the corresponding examples of plane domains.

We say that a point $e \in \tilde{D}$, $|e| \neq \emptyset$, is *simple* in $\mathbb{R}^m$ if its body $|e|$, with respect to $\mathbb{R}^m$, contains one point exactly. Observe that e is a prime end of the I-st type if and only if the point $e \in \tilde{D} \setminus D$ is simple in $\mathbb{R}^m$.

By j we denote the natural projection of a surface $D \subset \mathbb{R}^m$ onto $\mathbb{R}^m$. The following statement belongs to Luferenko and Suvorov [94].

Theorem 4.4   *Assume that sets D and $j(D)$ are precompact in $\tilde{D}$ and $\mathbb{R}^m$ respectively. The boundary $\tilde{D} \setminus D$ contains exceptionally simple points in $\mathbb{R}^m$ if and only if the natural projection $j : D \to \mathbb{R}^m$ is uniformly continuous on D with respect to the relative distance ρ.*

For the **proof** see [94] or prove independently.

Exercises 4.5   1) Study Piranian's article [145] on the distribution of prime ends of plane domains and prove corresponding results in the case of surfaces. 2) Spread the concept of the Caratheodory kernel convergence of a domains sequence to the case of surfaces (see Carathéodory [23], Suvorov [177,]), Ryazanov [159], Nasyrov [136]).

4.3 The Conformal Map T

Let D be a simply connected subdomain of $\mathbb{R}^2$ with a nonempty boundary ∂D. Assume that a surface $\Omega \in \text{Lip}_{\text{loc}}(D)$ is embedded into $\mathbb{R}^m$ with a locally Lipschitz mapping (3.14).

Let $G \subset \mathbb{R}^2$ be a simply connected domain different from $\mathbb{R}^2$. We consider a conformal mapping $T : G \to \Omega$ that is a homeomorphism with properties $T, T^{-1} \in W_{\text{loc}}^{1,2}$ and such that at every point of differentiability, T transforms infinitesimal circles in G into infinitesimal circles with respect to the surface Ω.

By Theorem 3.12 we conclude that $T = F \circ h$ where $F : B(0, R) \to \Omega$ is a canonical homomorphism and $h : G \to B(0, R)$ is conformal.

Fix a point $O' \in \Omega$. We put

$$O'' = T^{-1}(O'), \quad r(G) = \inf_{u \in \partial G} |u - O''|.$$

The following Theorem 4.6 is a generalization of the classic Carathéodory Theorem [24] and guarantees that the conformal mapping $T : G \to \Omega$ is continuously extendable up to the continuous map $\widetilde{T} : \widetilde{G} \to \widetilde{\Omega}$.

Theorem 4.6 *Let Ω be a simply connected and locally bi-Lipschitz surface in $\mathbb{R}^m$ given in the form (3.14). Let $T : G \to \Omega$ be a conformal mapping.*

Then for every pair of points $p, q \in G$, satisfying the condition

$$\rho(p, q; O'', G) < \min\left\{1, \frac{1}{16} r^4(G)\right\}, \tag{4.7}$$

it is fulfilled

$$\rho\left((T(p), T(q); O', \Omega\right) \le K \log^{-1/2} \frac{1}{\rho(p, q; O'', G)} \tag{4.8}$$

with

$$K = 2\sqrt{\pi}\,(\text{area}\,\Omega)^{1/2}.$$

Proof. We fix points $p, q \in G \backslash \{O''\}$ with properties (4.7). Let $\gamma \subset G \backslash \{O''\}$ be an arc such that

$$\text{length}\,(\gamma) < \rho(p, q; O'', G) + \epsilon$$

where $\epsilon > 0$ is a sufficiently small number.

We choose a point $\xi \in \gamma$ and consider the family of the circles

$$L_\tau = \{u \in \mathbb{R}^2 : |u - \xi| = \tau\}, \quad r < \tau < R \tag{4.9}$$

where

$$r = \rho(p, q; O'', G) + \epsilon, \quad R = \sqrt{\rho(p, q; O'', G)}.$$

By (4.7) we have

$$\rho(p, q; O'', G) < \sqrt{\rho(p, q; O'', G)} < 1,$$

and the family is nonempty for sufficiently small $\epsilon > 0$.

Every circle L_τ contains a subarc γ.

Now we consider two cases.

The first case. Suppose that

$$\text{dist}\,(\xi, \partial G) \leq \sqrt{\text{length}\,(\gamma)} \qquad (4.10)$$

where dist (a, B) is the Euclid distance from a point a to a set B.

Then we have

$$\sqrt{\rho(p, q; O'', G)} < |\xi - O''|, \qquad (4.11)$$

i.e. every L_τ separates p, q from O'.

Indeed, assume the contrary. Suppose that

$$|\xi - O''| \leq \sqrt{\rho(p, q; O'', G)}.$$

For a point

$$u \in \partial G \quad \text{such that} \quad |u - \xi| = \text{dist}\,(\xi, \partial G),$$

we have

$$|u - O''| \leq |u - \xi| + |\xi - O''|.$$

Thus by (4.10),

$$r(G) \leq |u - O''| \leq \sqrt{\text{length}\,(\gamma)} + \sqrt{\rho(p, q; O'', G)}$$

$$\leq \sqrt{\rho(p, q; O'', G) + \epsilon} + \sqrt{\rho(p, q; O'', G)}.$$

The relation (4.9) between r and R implies

$$r(G) \leq \rho^{1/4}(p, q; O'', G) + \rho^{1/2}(p, q; O'', G) \leq 2\,\rho^{1/4}(p, q; O'', G),$$

and

$$\frac{1}{16} r^4(G) \leq \rho(p, q; O'', G).$$

This contradicts with (4.7), and the inequality (4.11) is valid.

For arbitrary τ, $r \leq \tau \leq R$, let C_τ be a component of $L_\tau \cap G$, separating points p, q from O''. The existence of such component is a corollary of (4.11).

Let $\Delta_{r,R}$ be a subdomain of G, situated between C_r and C_R.

The map $T(u) : G \to \mathbb{R}^m$ belongs to $W^{1,2}_{\text{loc}}(G)$, and hence, it is absolutely continuous along a.e. arcs C_τ, $r < \tau < R$.

Let $u(s)$ be a natural parametrization of the arc C_τ. Here $0 \le s \le l(C_\tau)$, and $l(C_\tau)$ is the length of C_τ. Then

$$l(T(C_\tau)) = \int\limits_{C_\tau} \left| \frac{dT}{ds}(u(s)) \right| ds \le \int\limits_{C_\tau} \sqrt{\sum_{i=1}^{m} |\nabla y_i(u)|^2} \, |du|.$$

It is not difficult to prove, that the function $l(T(C_\tau))$ is Lebesgue measurable as a function of the parameter $\tau \in [r, R]$. By the Cauchy integral inequality, we have

$$l^2 (T(C_\tau)) \le l(C_\tau) \int\limits_{C_\tau} \sum_{i=1}^{m} |\nabla y_i(u)|^2 \, |du|,$$

and by Theorem 1.51 we obtain

$$\int\limits_{r}^{R} \frac{l^2 (u(C_\tau))}{\tau} \, d\tau \le 2\pi \int\limits_{\Delta_{r,R}} \sum_{i=1}^{m} |\nabla y_i(u)|^2 \, du_1 \, du_2. \tag{4.12}$$

We consider the conformal map T. At each point where $T : G \to \Omega$ is differentiable, we have

$$\sum_{i=1}^{m} |\nabla y_i(u)|^2 = \sum_{k=1}^{2} \left| \frac{\partial T}{\partial u_k} \right|^2 =$$

$$= 2 \left| \frac{\partial T}{\partial u_1} \right|^2 = 2\sqrt{\left| \frac{\partial T}{\partial u_1} \right|^2 \left| \frac{\partial T}{\partial u_2} \right|^2}.$$

Thus

$$\sum_{i=1}^{m} |\nabla y_i(u)|^2 = 2 \, I(T, u) \quad \text{a.e. on} \quad G \tag{4.13}$$

where

$$I(T, u) = \sqrt{\left| \frac{\partial T}{\partial u_1} \right|^2 \left| \frac{\partial T}{\partial u_2} \right|^2 - \left\langle \frac{\partial T}{\partial u_1}, \frac{\partial T}{\partial u_2} \right\rangle^2}.$$

Because the area of Ω is finite and the vector function $T : G \to \mathbb{R}^m$ belongs to $W^{1,2}_{\text{loc}}(G)$, the relation (4.13) implies

$$\int\limits_{G} \sum_{i=1}^{m} |\nabla y_i|^2 \, du_1 \, du_2 = 2 \, \text{area}\,(\Omega) < \infty.$$

Therefore, by (4.12) we arrive to the estimate

$$\inf_{r \le \tau \le R} l^2(T(C_\tau)) \le 2\pi \log^{-1} \frac{\sqrt{\rho(p,q;O'',G)}}{\rho(p,q;O'',G)+\epsilon} \int_G \sum_{i=1}^m |\nabla y_i(u)|^2 \, du_1 \, du_2. \quad (4.14)$$

By (4.14), taking into account the arbitrariness of $\epsilon > 0$, we conclude

$$\inf_{r \le \tau \le R} l(T(C_\tau)) \le \sqrt{4\pi \operatorname{area}(\Omega)} \log^{-1/2} \frac{1}{\rho(p,q;O'',G)}.$$

Since every arc C_τ separates points p, q from O'' on G, then every arc $T(C_\tau)$ separates $T(p)$, $T(q)$ from O' on Ω. Thus we obtain

$$\rho_2(T(p),T(q);O',F) \le K_1 \log^{-1/2} \frac{1}{\rho(p,q;O'',G)}.$$

This inequality implies the validity of (4.8).

The second case. Suppose that $\operatorname{dist}(\xi, \partial G) > \sqrt{\operatorname{length}(\gamma)}$. Then we have

$$\operatorname{dist}(\xi, \partial G) > \sqrt{\rho(p,q;O'',G)} = R,$$

and every circle C_τ of the family (4.9) does not intersect ∂G.

As above, we prove the estimate (4.14).

The mapping $T : G \to \Omega$ is homeomorphic, and hence, every curve $T(C_\tau)$, $r \le \tau \le R$, separates points $T(p)$, $T(q)$ from the boundary $\partial \Omega$ on Ω, and for every $\tau \in [r, R]$ it is fulfilled

$$\rho_1(T(p),T(q);O',\Omega) \le \operatorname{length}(T(C_\tau)) \le l(T(C_\tau)).$$

Thus by (4.14), we obtain

$$\rho_1(T(p),T(q);O',\Omega) \le K_1 \log^{-1/2} \frac{1}{\rho(p,q;O'',G)}.$$

This implies (4.8), and the theorem is proved.   □

Corollary 4.15   *Let Ω be a simply connected and locally bi-Lipschitz surface in $\mathbb{R}^m$ given in the form (3.14). Assume that the function $\Lambda^2(x)/\lambda^2(x)$ is $W^{1,2}$-majorized on D.*

Then every conformal mapping $T : G \to \Omega$ where $G \subset \mathbb{R}^2$ is a simply connected domain with a nonempty boundary, is continuously extended up to a continuous mapping $\tilde{T} : \tilde{G} \to \tilde{\Omega}$.

4.4 Q^*-homeomorphisms of Surfaces

Let D be a simply connected subdomain of $\mathbb{R}^2$ with a nonempty boundary ∂D. Suppose that a surface Ω is embedded into $\mathbb{R}^m$ by a locally bi-Lipschitz mapping (3.14).

Let $U, V \subset D$ be nonintersecting subsets closed with respect to D. The triple $(U, V; D)$ defines a *condenser* on the surface Ω. We consider the set $\mathcal{F}(U, V; D)$ of all Lipschitz functions $\varphi : D \to \mathbb{R}$ such that $\varphi|_U = 0$, $\varphi|_V = 1$ and

$$|\nabla_\Omega \varphi(x)| > 0 \quad \text{a.e. on} \quad D \setminus (U \cup V). \tag{4.16}$$

Here the gradient $\nabla_\Omega \varphi$ is calculated with respect to the metric of the surface Ω.

The quantity

$$\text{cap}_1 (U, V; D) = \inf_{\varphi \in \mathcal{F}(U,V;D)} \int_D |\nabla_\Omega \varphi| \, d\sigma_\Omega. \tag{4.17}$$

is the 1-capacity of the condenser $(U, V; D)$.

Let $\Omega_1 \subset \mathbb{R}^m$, $\Omega_2 \subset \mathbb{R}^n$ be locally bi-Lipschitz surfaces of the form (3.14). Let $D_1 \subset \Omega_1$, $D_2 \subset \Omega_2$ be domains and let $h : D_1 \to D_2$ be a homeomorphic mapping of D_1 onto D_2. Fix a measurable function $Q : D_1 \to (0, \infty)$. A homeomorphic map h is called the Q^*-*homeomorphism* if, for every condenser $(U, V; D_1)$, it is fulfilled

$$\text{cap}_1^2 (hU, hV; D_2) \le \inf_{\varphi \in \mathcal{F}(U,V;D_1)} \int_{D_1} Q \, |\nabla_{\Omega_1} \varphi|^2 \, d\sigma_{\Omega_1}. \tag{4.18}$$

For $Q \equiv \text{mes}_2 D_2$ the relation (4.18) *follows from the conformality of* h. Indeed, let $z = x_1 + ix_2$, $w = u_1 + iu_2$ and let $w = h(z) : D_1 \to D_2$ be the schlicht conformal mapping between domains $D_1, D_2 \subset \mathbb{C}$. The conformal map h leaves invariant the Dirichlét integral that is

$$\int_{D_2} |\nabla \varphi(w)|^2 du_1 du_2 = \int_{D_1} |\nabla \varphi^*(z)|^2 dx_1 dx_2 \tag{4.19}$$

where $\varphi^* = \varphi \circ h$. Hence, for every condenser $(U, V; D_2)$ and an arbitrary function $\varphi \in \mathcal{F}(U, V; D_2)$, we have

$$\mathrm{cap}_1^2(U, V; D_2) \leq \left(\int\limits_{D_2} |\nabla \varphi(w)| \, du_1 du_2 \right)^2 \leq$$

$$\leq \mathrm{mes}_2 \, D_2 \int\limits_{D_2} |\nabla \varphi(w)|^2 \, du_1 du_2 =$$

$$= \mathrm{mes}_2 \, D_2 \int\limits_{D_1} |\nabla \varphi^*(z)|^2 \, dx_1 dx_2 \,.$$

Passing on to the infimum over all functions

$$\varphi^* \in \mathcal{F}(h^{-1}U, h^{-1}V; D_1) \,,$$

we arrive to the relation (4.18).

In just the same way, we check that for $Q \equiv \mathrm{const} \geq 0$ the class of Q^*-homeomorphisms $h : D_1 \subset \mathbf{C} \to \mathbf{C}$ contains the quasi-conformal mappings.

In the general case the inequality (4.18) can be interpreted as a special variant of the Length and Area Principle for the map h (see, for example, [91, Theorem **1.4.b**], [179, Chapter X, Theorem 1], [75], [79], [73]). In the case of subdomains D_1, D_2 of $\mathbb{R}^2$, this class is connected with Q-homeomorphic mappings which were introduced by Strugov [175] and studied by Martio, Ryazanov, Srebro, Yakubov [101] – [102], and also radial Q-homeomorphisms [55], [160].

In the considered case of Lipschitz surfaces, we are right to assume that the mapping h is conformal only at the points where h is differentiable, or that $h : D_1 \to D_2$ is conformal a.e. on D_1. On the other hand, we obtain from the assumption (4.19) of the Dirichlét integral invariance a sufficiently effective definition of the conformality of h. As above, we can check that such maps belong to the class of Q^*-homeomorphisms with $Q \equiv \mathrm{const} > 0$.

Example 4.20   Consider another example. Let $\Omega_1 = \Omega_2 \subset \mathbb{R}^2$ and let $\sigma : D_1 \to \mathbb{R}$ be a positive measurable function. We consider the system

$$u_{x_1} = \sigma \, v_{x_2}, \quad u_{x_2} = -\sigma \, v_{x_1}. \tag{4.21}$$

The following statement holds.

Theorem 4.22 *Each homeomorphic map* $h = u + iv : D_1 \to D_2$, $h \in W_{\text{loc}}^{1,2}$, *is* Q_1^*-*homeomorphic with*

$$Q_1 = Q(x) \operatorname{mes}_2 D_2$$

where

$$Q(x) = \max\left(\sigma(x), \frac{1}{\sigma(x)}\right).$$

Proof. Indeed, let $(U, V; D_2)$ be a condenser and $\varphi \in \mathcal{F}(U, V; D_2)$. We have

$$\operatorname{cap}_1^2 (U, V; D_2) \leq \operatorname{mes}_2 D_2 \int_{D_2} |\nabla \varphi|^2 du_1 du_2.$$

By Rademacher's Theorem, the function φ is differentiable a.e. on the domain D_2, and by the Gehring-Lehto theorem [42], the map h is also differentiable a.e. on D_1. Thus we are right to assume that

$$\int_{D_2} |\nabla \varphi|^2 du_1 du_2 = \int_{D_1} (\varphi_{u_1}^2 + \varphi_{u_2}^2)(u_{1x_1} u_{2x_2} - u_{1x_2} u_{2x_1}) \, dx_1 dx_2.$$

Using (4.21), we obtain

$$\operatorname{cap}_1^2 (U, V; D_2) \leq \operatorname{mes}_2 D_2 \int_{D_1} \sigma (\varphi_{u_1}^2 + \varphi_{u_2}^2) |\nabla u_2|^2 \, dx_1 dx_2.$$

By the change of variables formula (see, [56, Theorem **11**]), we conclude that the function $\varphi^* = \varphi \circ h$ is differentiable a.e. on D_1. Thus

$$(\varphi^*)_{x_1}^2 + (\varphi^*)_{x_2}^2 = (\sigma \varphi_{u_1} u_{2x_2} + \varphi_{u_2} u_{2x_1})^2 + (-\sigma \varphi_{u_1} u_{2x_1} + \varphi_{u_2} u_{2x_2})^2 =$$

$$= (\sigma^2 \varphi_{u_1}^2 + \varphi_{u_2}^2) |\nabla u_2|^2.$$

However, a.e. on D_1 it is fulfilled

$$(\sigma^2 \varphi_{u_1}^2 + \varphi_{u_2}^2) \leq \sigma Q (\varphi_{u_1}^2 + \varphi_{u_2}^2),$$

and hence,

$$\operatorname{cap}_1^2 (U, V; D_2) \leq \operatorname{mes}_2 D_2 \int_{D_1} Q ((\varphi^*)_{x_1}^2 + (\varphi^*)_{x_2}^2) \, dx_1 dx_2$$

that proves (4.18). $\square$

The solutions of (4.21) are called σ-*harmonic mappings* and studied by many authors (see, for example, [9], [6], [36] and references).

□

Let $\Omega_1 \subset \mathbb{R}^m$ be a locally bi-Lipschitz surface of the form (3.14). Denote by $d_{\Omega_1}(a, b) = d(a, b)$, the geodesic distance between points a, $b \in \Omega_1$. In other words, $d_{\Omega_1}(a, b)$ is the infimum over the lengths of arcs $\gamma \subset \Omega_1$, joining points a and b. Because the surface Ω_1 is locally bi-Lipschitz, then $d_{\Omega_1}(a, b) < \infty$ for every $a, b \in \Omega_1$. Other designations are:

$$S(y_0, r) = \{y \in \Omega_1 : d_{\Omega_1}(y_0, y) = r\}$$

for a geodesic circle of the radius $r > 0$,

$$B(y_0, r) = \{y \in \Omega_1 : d_{\Omega_1}(y_0, y) < r\}$$

for a geodesic disc, and

$$K(y_0, r, R) = \{y \in \Omega_1 : r < d_{\Omega_1}(y_0, y) < R\}$$

for a geodesic ring.

Fix a simply connected domain $D \subset \Omega_1$ and a point $y_0 \in \overline{D}$. Choose r and R such that

$$0 < r < R < \sup_{y \in D} d_{\Omega_1}(y_0, y), \tag{4.23}$$

and a component $K = K_D(y_0, r, R)$ of the set $K(y_0, r, R) \cap D$. We put

$$U = U(y_0, r) = D \cap B(y_0, r), \quad V = V(y_0, R) = D \setminus B(y_0, R).$$

Since $y_0 \in \overline{D}$, then from (4.23), it follows that the sets U and V are nonempty.

Theorem 4.24 *Let $\Omega_1 \subset \mathbb{R}^m$, $\Omega_2 \subset \mathbb{R}^n$ be locally bi-Lipschitz surfaces of the form (3.14). If*

$$h : K_D(y_0, r, R) \subset \Omega_1 \to \Omega_2$$

is a Q^*-homeomorphism, then

$$L^2 \leq \left(\int_r^R dt \bigg/ \int_{S_D(y_0,t)} Q(y)\, |dy| \right)^{-1}. \tag{4.25}$$

Here L is the infimum over lengths of arcs or curves, separating sets hU, hV on hK, and $S_D(y_0, t) = S(y_0, t) \cap D$.

The proof of this theorem see below. At first, we formulate some of its important corollaries.

Corollary 4.26 *Under assumptions of Theorem 4.24 let y_0 be a principal point of a prime end $e \in \tilde{D}$. If*

$$\int\limits_0^t \frac{dt}{\displaystyle\int\limits_{S_D(y_0,t)} Q(y)\,|dy|} = \infty, \tag{4.27}$$

then the image of $e \in \tilde{D}$ under a map $h : K_D(y_0, 0, R) \to \Omega_2$ is a unique prime end of the domain

$$h(K_D(y_0, 0, R)) \subset D_2\,.$$

Proof. Let $a_i \to e$ be a sequence of points of the domain $K_D(y_0, 0, R)$ $(i = 1, 2, \ldots)$. It is sufficient to observe that relations (4.25), (4.27) imply the existence of a sequence $\{h(S_D(y_0, t_k))\}$ of arcs with lengths, tending to zero as $k \to \infty$ and separating the arc $h(S_D(y_0, R))$ from $h(a_i)$ for all sufficiently large i. $\qquad\qquad\square$

Along with Q^*-homeomorphisms, we consider their generalizations. Namely, fix a simply connected domain $D_1 \subset \Omega_1$ and a point $O_1 \in D_1$. Let

$$0 < \mu < d(O_1) = d(O_1, \partial D_1) \tag{4.28}$$

be a number. We say that a homeomorphic map

$$h : D_1 \subset \Omega_1 \to D_2 \subset \Omega_2$$

is a $Q^*(\mu)$-*homeomorphism* if, for every point $y_0 \in \overline{D_1}$ and for arbitrary $0 < r < R < \mu$, there exists a measurable function $Q = Q(y; y_0, r, R)$ such that

$$\mathrm{cap}_1^2(hU(y_0, r),\, hV(y_0, R); D_2) \le \inf_\varphi \int\limits_{K_{D_1}(y_0,r,R)} Q\,|\nabla_{\Omega_1}\varphi|^2 d\sigma_{\Omega_1} \tag{4.29}$$

where the infimum is taken over all functions

$$\varphi \in \mathcal{F}(U(y_0, r),\, V(y_0, R);\, K_{D_1}(y_0, r, R)).$$

Theorem 4.30 *Let $\Omega_1 \subset \mathbb{R}^m$, $\Omega_2 \subset \mathbb{R}^n$ be a locally bi-Lipschitz surfaces of the form (3.14). Let $D_1 \subset \Omega_1$, $D_2 \subset \Omega_2$ be simply connected domains and $O_1 \in D_1$, $O_2 \in D_2$ be fixed points. Let $h : D_1 \to D_2$ be a $Q^*(\mu)$-homeomorphism, $h(O_1) = O_2$ such that for a monotone function $\omega(t) = \omega(t; \mu)$:*

$(0,\mu) \to (0,\infty)$, $\omega(+0) = 0$, *an arbitrary point* $y_0 \in \overline{D}_1$, *and every* $0 < t < \mu$, *it is fulfilled*

$$\left(\int\limits_{t}^{\mu} d\tau \Big/ \int\limits_{S_{D_1}(y_0,\tau)} Q(y)\,|dy| \right)^{-1} \leq \omega(t). \qquad (4.31)$$

If $\mathrm{mes}_2\, D_2 < \infty$ *then for every pair of points* $a,b \in \tilde{D}_1$, *satisfying the condition*

$$2\max\{d(a,\partial D_1), \quad d(b,\partial D_1), \rho(a,b;O_1,D_1)\} <$$

$$< \min\{d(O_1,\partial D_1) - \mu, 2\mu\}, \qquad (4.32)$$

the following inequality holds

$$\rho(ha,\, hb; O_2, D_2) \leq [\mathrm{mes}_2\, D_2\, \omega(\rho(a,b;O_1,D_1))]^{1/2}. \qquad (4.33)$$

In particular, every $Q^*(\mu)$-homeomorphism $h : D_1 \to D_2$ with the property (4.31) is continuously extended up to a homeomorphic mapping $\tilde{h} : \tilde{D}_1 \to \tilde{D}_2$, and we obtain a generalization of Carathéodory Theorem on conformal mappings of plane domains [24] up to the case of mappings between surfaces.

Corollary 4.34   *If a* $Q^*(\mu)$-*homeomorphism* $h : D_1 \to D_2$ *satisfies* (4.31), *and all prime ends of the surface* D_2 *have the first type, then the mapping* h *is continuously extended up to the homeomorphic mapping* $\tilde{h} : \tilde{D}_1 \to \overline{D}_2$.
     In addition, if the surface D_1 *has the prime ends of the first type, then* h *continuously extended up to the homeomorphism* $\tilde{h} : \overline{D}_1 \to \overline{D}_2$.

4.5   Proof of Theorem 4.24

At first, we will prove that

$$L \leq \mathrm{cap}_1\,(hU, hV; hK). \qquad (4.35)$$

Let φ be a function, admissible under calculations of the 1-capacity of the condenser $(hU, hV; hK)$. Suppose that the surface $\Omega_2 \subset \mathbb{R}^n$ is given by a Lipschitz mapping $f_2 : \Delta^* \to \mathbb{R}^n$. Let $K^* = f_2^{-1}(hK)$. We are right to assume that

$$\iint\limits_{hK} |\nabla_{\Omega_2} \varphi|\, d\sigma_{\Omega_2} = \iint\limits_{K^*} \left(\sum_{i,j=1}^{2} g^{ij} \varphi_{x_i}^* \varphi_{x_j}^* \right)^{1/2} \sqrt{g}\, dx_1 dx_2 \qquad (4.36)$$

where $\varphi^* = \varphi \circ f_2$, and g_{ij}, g^{ij}, g are coefficients of the first quadratic form of the surface Ω_2.

Since the mapping f_2 is locally bi-Lipschitz, then

$$\operatorname*{ess\,sup}_{K^*} \max\{g_{11}(x), |g_{12}(x)|, g_{22}(x)\} = M < \infty.$$

We put

$$p(x) = \left(\sum_{i,j=1}^{2} g^{ij} \frac{\varphi^*_{x_i}}{|\nabla\varphi^*|} \frac{\varphi^*_{x_j}}{|\nabla\varphi^*|} \right)^{1/2} \sqrt{g}.$$

Then, we have

$$\iint\limits_{hK} |\nabla_{\Omega_2}\varphi|\, d\sigma_{\Omega_2} = \iint\limits_{K^*} p(x)\, |\nabla\varphi^*(x)|\, dx_1 dx_2.$$

Because matrices (g_{ij}) and (g^{ij}) are mutually inverse,

$$g^{11} = \frac{g_{22}}{g}, \quad g^{12} = -\frac{g_{12}}{g}, \quad g^{22} = \frac{g_{11}}{g}. \tag{4.37}$$

Thus, a.e. on K^*

$$p(x) \le M^{1/2} \left(\sum_{i,j=1}^{2} \frac{\varphi^*_{x_i}}{|\nabla\varphi^*|} \frac{\varphi^*_{x_j}}{|\nabla\varphi^*|} \right)^{1/2} \le 2\, M^{1/2}.$$

By [37, Theorem **3.2.15**], a.e. level sets

$$E_t = \{x \in K^* : \varphi^*(x) = t\}$$

consist of the countable number of locally rectifiable arcs. Hence, we are right to use the well-known co-area formula [37, Theorem **3.2.22**]. We have

$$\iint\limits_{K^*} \left(\sum_{i,j=1}^{2} g^{ij} \varphi^*_{x_i} \varphi^*_{x_j} \right)^{1/2} \sqrt{g}\, dx_1 dx_2 =$$

$$= \int_0^1 dt \int\limits_{E_t} \left(\sum_{i,j=1}^{2} g^{ij} \frac{\varphi^*_{x_i}}{|\nabla\varphi^*|} \frac{\varphi^*_{x_j}}{|\nabla\varphi^*|} \right)^{1/2} \sqrt{g}\, |dx|. \tag{4.38}$$

The vector function f_2 is absolutely continuous along a.e. E_t. The vectors

$$\overline{\tau} = \left(-\frac{\varphi^*_{x_2}}{|\nabla \varphi^*|}, \frac{\varphi^*_{x_1}}{|\nabla \varphi^*|} \right)$$

are unit vectors tangent to E_t. Thus, for a.e. $t \in (0, 1)$, we obtain

$$\text{length}(f_2 E_t) \;\; = \int\limits_{E_t} ds = \int\limits_{E_t} \left(\sum_{i,j=1}^{2} g_{ij} dx_i dx_j \right)^{1/2} =$$

$$= \int\limits_{E_t} \left(g_{11} \frac{(\varphi^*_{x_2})^2}{|\nabla \varphi^*|^2} - 2g_{12} \frac{\varphi^*_{x_1}}{|\nabla \varphi^*|} \frac{\varphi^*_{x_2}}{|\nabla \varphi^*|} + g_{22} \frac{(\varphi^*_{x_1})^2}{|\nabla \varphi^*|^2} \right)^{1/2} |dx|.$$

Substituting quantities of g^{ij} from (4.37) to the last integral, we have

$$\text{length}(f_2 E_t) = \int\limits_{E_t} \left(\sum_{i,j=1}^{2} g^{ij} \frac{\varphi^*_{x_i}}{|\nabla \varphi^*|} \frac{\varphi^*_{x_j}}{|\nabla \varphi^*|} \right)^{1/2} \sqrt{g} \, |dx|. \qquad (4.39)$$

Further (4.38) implies

$$\int\limits_0^1 \text{length}(f_2 E_t) \, dt = \iint\limits_{K^*} \left(\sum_{i,j=1}^{2} g^{ij} \varphi^*_{x_i} \varphi^*_{x_j} \right)^{1/2} \sqrt{g} dx_1 dx_2.$$

Therefore,

$$\inf_{0<t<1} \text{length}(f_2 E_t) \le \int\limits_{K^*} \left(\sum_{i,j=1}^{2} g^{ij} \varphi^*_{x_i} \varphi^*_{x_j} \right)^{1/2} \sqrt{g} \, dx_1 dx_2.$$

Using (4.36), we arrive to (4.35).

As the second step, we prove that

$$\inf_{\varphi \in \mathcal{F}(U,V,D)} \iint\limits_{K} Q |\nabla_{\Omega_1} \varphi|^2 d\sigma_{\Omega_1} \le \left(\int\limits_r^R dt \Big/ \int\limits_{S_D(y_0,t)} Q(y) \, |dy| \right)^{-1}. \qquad (4.40)$$

Suppose that the surface $\Omega_1 \subset \mathbb{R}^m$ is given by a locally bi-Lipschitz mapping f_1. As above, let $K^* = f_1^{-1}(K)$, $\varphi^* = \varphi \circ f_1$ and $d^*(x) = d_{\Omega_1}(y_0, f_1(x))$. We have

$$\inf_{\varphi \in \mathcal{F}(U,V;D)} \iint_K Q\,|\nabla_{\Omega_1}\varphi|^2 d\Omega_1 = \inf_{\varphi^*} \iint_{K^*} Q^* \sum_{i,j=1}^{2} g^{ij} \varphi^*_{x_i} \varphi^*_{x_j} \sqrt{g}\, dx_1 dx_2$$

(4.41)

where $Q^* = Q \circ f_1$ and g^{ij}, g are coefficients of the first quadratic form of the surface Ω_1.

We choose φ^* in the form $\varphi^* = \psi \circ d^*(x)$ with a Lipschitz function $\psi : [r, R] \to [0, 1]$ such that $\psi(r) = 0$, $\psi(R) = 1$. We will need the following estimate for the gradient of the distance function which we will be proved below:

$$|\nabla_{F_1} d^*(x)| = \sum_{ij=1}^{2} g^{ij} d^*_{x_i} d^*_{x_j} \le 1 \quad \text{a.e. on} \quad \Delta.$$

(4.42)

Thus, using again the co-area formula by (4.42), we find

$$\iint_{K^*} Q^* \sum_{i,j=1}^{2} g^{ij} \varphi^*_{x_i} \varphi^*_{x_j} \sqrt{g}\, dx_1 dx_2 =$$

$$= \iint_{K^*} Q^* \psi'^2(d^*(x)) \sum_{i,j=1}^{2} g^{ij} d^*_{x_i} d^*_{x_j} \sqrt{g}\, dx_1 dx_2 \le$$

(4.43)

$$\le \int_r^R \psi'^2(t) dt \int_{S^*_D(y_0,t)} Q^* \sqrt{g} |dx|$$

where $S^*_D(y_0, t) = f_1^{-1} S_D(y_0, t)$.

It is easy to see that

$$\int_{S^*_D(y_0,t)} Q^* \sqrt{g}\, |dx| = \int_{S_D(y_0,t)} Q(y)\, |dy|.$$

(4.44)

Indeed, we have

$$\int_{S_D(y_0,t)} Q(y)\, |dy| = \int_{S^*_D(y_0,t)} Q^* \left(\sum_{i,j=1}^{2} g_{ij} dx_i dx_j \right)^{1/2}.$$

Since $d^*|_{S_D^*(y_0,t)} = t$, as in (4.39), we establish that for a.e. $t \in (0,1)$

$$\int_{S_D^*(y_0,t)} Q^* \left(\sum_{i,j=1}^{2} g_{ij} dx_i dx_j \right)^{1/2} = \int_{S_D^*(y_0,t)} Q^* \left(\sum_{i,j=1}^{2} g^{ij} d_{x_i}^* d_{x_j}^* \right)^{1/2} \sqrt{g}\, |dx| =$$

$$= \int_{S_D^*(y_0,t)} Q^* \sqrt{g}\, |dx|,$$

and (4.44) is proved.

From (4.43) and (4.44), it follows

$$\iint_{K^*} Q^* \sum_{i,j=1}^{2} g^{ij} \varphi_{x_i}^* \varphi_{x_j}^* \sqrt{g}\, dx_1 dx_2 \leq \int_r^R \psi'^2(t) dt \int_{S_D(y_0,t)} Q(y)\, |dy|.$$

Thus, (4.41) implies

$$\inf_{\varphi \in \mathcal{F}(U,V;D)} \iint_K Q |\nabla_{\Omega_1} \varphi|^2 d\sigma_{\Omega_1} \leq \inf_{\psi} \int_r^R \psi'^2(t) dt \int_{S_D(y_0,t)} Q(y)\, |dy|. \qquad (4.45)$$

Now, we will find the infimum over integrals at the right side of (4.45). Since $\psi(r) = 0$, $\psi(R) = 1$, then for a function ψ we have

$$1 \leq \left(\int_r^R \frac{dt}{\int_{S_D(y_0,t)} Q(y)\, |dy|} \right) \left(\int_r^R \psi'^2(t)\, dt \int_{S_D(y_0,t)} Q(y)\, |dy| \right),$$

that is, for every admissible function ψ, the following inequality holds

$$\left(\int_r^R \frac{dt}{\int_{S_D(y_0,t)} Q(y)\, |dy|} \right)^{-1} \leq \int_r^R \psi'^2(t)\, dt \bigg/ \int_{S_D(y_0,t)} Q(y)\, |dy|. \qquad (4.46)$$

Choosing

$$\psi_0(t) = \int_r^t \frac{d\tau}{\int_{S_D(y_0,\tau)} Q(y)\, |dy|} \bigg/ \int_r^R \frac{d\tau}{\int_{S_D(y_0,\tau)} Q(y)\, |dy|}$$

for $t \in (r, R)$, and observing that $\psi_0(r+0) = 0$, $\psi_0(R-0) = 1$, we conclude:

$$\int_r^R \psi_0'^2(t)\, dt \int_{S_D(y_0,t)} Q(y)\, |dy| = \left(\int_r^R dt \Big/ \int_{S_D(y_0,t)} Q(y)\, |dy| \right)^{-1}.$$

From this, using (4.46), we find

$$\inf_\psi \int_r^R \psi'^2(t)\, dt \int_{S_D(y_0,t)} Q(y)\, |dy| =$$

$$= \left(\int_r^R dt \Big/ \int_{S_D(y_0,t)} Q(y)\, |dy| \right)^{-1}. \qquad (4.47)$$

These arguments are suitable for every function, locally bounded on (r, R),

$$q(\tau) \equiv \int_{S_D(y_0,\tau)} Q(y)\, |dy|$$

for which ψ_0 is locally Lipschitz. Simple approximative arguments imply the validity of (4.47) in the general case.

Indeed, for every $n = 1, 2, \ldots$, we put

$$Q_n(y) = \begin{cases} Q(y) & \text{for} \quad Q(y) \le n, \\ n & \text{for} \quad Q(y) > n. \end{cases}$$

Then for a function ψ by Fatou's theorem, we have

$$\int_r^R \psi'^2(t)\, dt \int_{S_D(y_0,t)} Q(y)\, |dy| \le \underline{\lim}_{n \to \infty} \int_r^R \psi'^2(t)\, dt \int_{S_D(y_0,t)} Q_n(y)\, |dy|.$$

From this, by the proved above,

$$\int\limits_{r}^{R} \psi'^{2}(t)\,dt \int\limits_{S_D(y_0,t)} Q(y)\,|dy| \le \int\limits_{r}^{R} \psi'^{2}(t)\,dt \int\limits_{S_D(y_0,t)} Q_n(y)\,|dy| \le$$

$$\le \left(\int\limits_{r}^{R} dt \Bigg/ \int\limits_{S_D(y_0,t)} Q_n(y)\,|dy| \right)^{-1} \le$$

$$\le \left(\int\limits_{r}^{R} dt \Bigg/ \int\limits_{S_D(y_0,t)} Q(y)\,|dy| \right)^{-1}.$$

Passing onto the infimum over all admissible functions ψ, we verify that (4.47) holds in the general case.

Combining (4.47) with (4.45), we arrive to (4.40). By (4.40), (4.35) and (4.18), we have (4.25).

To complete the proof of Theorem 4.24, it is necessary to check the validity of the inequality (4.42) used before. If the surface Ω_1 belongs to C^2 then this inequality is well-known (and so in the stronger form). For Lipschitz surfaces it is required to prove.

Let a be a point in which the derivative f_1' of the vector function (3.14) exists and is continuous. Let $\gamma(a,\theta)$ be a segment of the length $h > 0$, outgoing from a in a direction, given by the unit vector $\theta = (\theta_1, \theta_2)$. Then,

$$\left| \sum_{i=1}^{2} \frac{\partial d^*}{\partial x_i}(a)\,\theta_i \right| \le \varliminf_{h \to 0} \frac{1}{h} \int\limits_{\gamma(a,\theta)} ds_{\Omega_1}$$

where

$$d^*(x) = d(a, f_1(x)) = \inf_{\gamma} \int\limits_{\gamma} ds_{\Omega_1},$$

and the infimum is taken over all arcs γ joining points a and x.

However,

$$\int\limits_{\gamma(a,\theta)} ds_{F_1} = \int\limits_{0}^{h} \left[\sum_{i,j=1}^{2} g_{ij}(a + r\,\theta)\,\theta_i\theta_j \right]^{1/2} dr,$$

and by the continuity of g_{ij} at a, we obtain

$$\sum_{i,j=1}^{2} \frac{\partial d^*}{\partial x_i}(a) \frac{\partial d^*}{\partial x_j}(a)\, \theta_i \theta_j \leq \sum_{i,j=1}^{2} g_{ij}(a)\, \theta_i \theta_j \,. \tag{4.48}$$

It is easy to check that

$$\max_{|\theta|=1} \frac{\sum_{i,j=1}^{2} \dfrac{\partial d^*}{\partial x_i} \dfrac{\partial d^*}{\partial x_j} \theta_i \theta_j}{\sum_{i,j=1}^{2} g_{ij}\, \theta_i \theta_j} = \sum_{i,j=1}^{2} g^{ij} \frac{\partial d^*}{\partial x_i} \frac{\partial d^*}{\partial x_j} \,. \tag{4.49}$$

Indeed, by virtue of well-known properties of quadratic forms (see, for example, [41, pp. 289-290]), the maximum of the ratio of two quadratic forms in the left side of (4.49) equals to the maximal root λ of the equation

$$\det\left(\frac{\partial d^*}{\partial x_i} \frac{\partial d^*}{\partial x_j} - \lambda\, g_{ij} \right) = 0\,.$$

Multiplying both sides of this equality by $\det(g^{ij})$, we obtain

$$\det\left(a_{ij} - \lambda\, \delta_{ij} \right) = 0$$

where

$$a_{ij} = \frac{\partial d^*}{\partial x_j} \sum_{k=1}^{2} g^{ik} \frac{\partial d^*}{\partial x_k}$$

and δ_{ij} is the Kronecker symbol.

From this we find

$$\lambda_{\max} = \sum_{i=1}^{2} \sum_{j=1}^{2} g^{ij} \frac{\partial d^*}{\partial x_j} \sum_{k=1}^{2} g^{ik} \frac{\partial d^*}{\partial x_k}$$

that is necessary.

By (4.49) validity of (4.48) for all unit vectors θ, implies (4.42).

Thus the theorem is proved completely. $\qquad\qquad\square$

4.6   Local Estimates

Theorem 4.24 implies some local estimates for homeomorphic mappings of the class Q^*. Let $c \in [1, \infty)$ be a constant. A vector function $f : \Delta \to \mathbb{R}^m$ is called *c-monotone* if for every subdomain Δ', $\Delta' \Subset \Delta$, the following inequality holds

$$\operatorname{osc}(f, \Delta') \leq c\operatorname{osc}(f, \partial\Delta')\,. \tag{4.50}$$

We observe, for example, that a C^2-vector function f which defines a surface Ω of non-positive Gauss curvature, is c-monotone for $c = 1$. If a surface Ω has the convex graph in $\mathbb{R}^3$, then it is c-monotone with a constant $c = c(\Delta') \geq 1$ on every subdomain $\Delta' \subset\subset \Delta$.

Corollary 4.51   *Let $\Omega_1 \subset \mathbb{R}^m$, $\Omega_2 \subset \mathbb{R}^n$ be surfaces given by vector functions $f_1 : \Delta_1 \to \mathbb{R}^m$, $f_2 : \Delta_2 \to \mathbb{R}^n$ respectively. Assume that f_2 is c_2-monotone.*
   Then for a Q^*-homeomorphic mapping $h : \Omega_1 \to \Omega_2$, a point $y_0 \in \Omega_1$, and every r such that $c_2 r < d_0 = d_{\Omega_1}(y_0, \partial\Omega_1)$, the following inequality holds

$$\mathrm{osc}^2(h, B(y_0, r)) \leq c_2^2 \left(\int\limits_r^{d_0} dt \left/ \int\limits_{S(y_0,t)} Q(y)\, |dy| \right. \right)^{-1}. \tag{4.52}$$

For the **proof** it is sufficient to use (4.25) and observe that

$$\mathrm{osc}^2(h, B(y_0, r)) \leq c_2^2 L^2.$$

$\square$

The estimate (4.52) implies a series of well-known estimates. We restrict ourselves to the following construction. Let $\xi(t)$ be a nonnegative Borel function on $(0, d_0)$ such that

$$I(\varepsilon) = \int\limits_\varepsilon^{d_0} \xi(t)\, dt < \infty, \quad \varepsilon \in (0, d_0). \tag{4.53}$$

By the co-area formula, we have

$$\iint\limits_{\varepsilon < d_{\Omega_1}(y_0,y) < d_0} Q(y)\, \xi^2(d_{F_1}(y_0, y))\, d\sigma_{\Omega_1} = \int\limits_\varepsilon^{d_0} \xi^2(\tau)\, d\tau \int\limits_{S(y_0,\tau)} Q(y)\, |dy|.$$

From this, by the Cauchy inequality, we obtain

$$I^2(\varepsilon) \leq \left(\int\limits_\varepsilon^{d_0} d\tau \left/ \int\limits_{S(y_0,\tau)} Q(y)\, |dy| \right. \right) \left(\int\limits_\varepsilon^{d_0} \xi^2(\tau)\, d\tau \int\limits_{S(y_0,\tau)} Q(y)\, |dy| \right),$$

and

$$\left(\int\limits_\varepsilon^{d_0} d\tau \left/ \int\limits_{S(y_0,\tau)} Q(y)\, |dy| \right. \right)^{-1} \leq \frac{1}{I^2(\varepsilon)} \iint\limits_{\varepsilon < d_{F_1}(y_0,y) < d_0} Q(y)\, \xi^2(d_{F_1}(y_0, y))\, d\sigma_{\Omega_1}.$$

Thus the estimate (4.52) leads to the inequality

$$\operatorname{osc}^2(h, B(y_0, r)) \leq c_2^2 \, I^{-2}(r) \iint\limits_{K(r,d_0;F_1)} Q(y) \, \xi^2(d_{F_1}(y_0, y)) \, d\sigma_{\Omega_1} \qquad (4.54)$$

which is valid for every function ξ with the property (4.53).

The close inequality for maps of plane domains into the unit sphere was obtained by Ryazanov, Srebro and Yakubov [160].

4.7 Proof of Theorem 4.30

Let $a, b \in D_1 \setminus \{O_1\}$ be a pair of points satisfying (4.32). Fix an arc or a curve $\gamma \subset D_1 \setminus \{O_1\}$ separating a, b from O_1 and such that

$$\operatorname{length}(\gamma) < \rho(a, b; O_1, D_1) + \delta < \frac{1}{2}(d(O_1) - \mu)$$

where $d(O_1) = d(O_1, \partial D_1)$ and $\delta > 0$ is sufficiently small number.

Choose a point $p \in \gamma$ and consider the family of geodesic circles

$$S(p, \tau) = \{y \in \Omega_1 : d(p, y) = \tau\}, \quad \rho < \tau < \mu$$

where $\rho = \rho(a, b; O_1, D_1)$.

Assumptions (4.28) and (4.32) imply that the sets $S(p, \tau) \cap D_1$ contain components $S_D(p, \tau)$ separating a, b from O_1.

Indeed, let $S(p, \tau) \cap \partial D_1 \neq \emptyset$. Since $p \in \gamma$, then

$$d(p, \partial D_1) \leq \operatorname{length}(\gamma) < \frac{1}{2}(d(O_1) - \mu).$$

Now we observe that

$$d(O_1) \leq d(p, \partial D_1) + d(p, O_1) \leq$$

$$\leq \operatorname{length}(\gamma) + d(p, O_1) \leq$$

$$\leq (d(O_1) - \mu)/2 + d(p, O_1).$$

Thus

$$\frac{1}{2} d(O_1) + \frac{1}{2}\mu \leq d(p, O_1),$$

and hence,

$$d(p, \partial D_1) + \mu \leq d(p, O_1).$$

From this $\mu \leq d(p, O_1)$, that is $\tau < d(p, O_1)$ and every arc $S_{D_1}(p, \tau) \neq \emptyset$.

Let $S(p, \tau) \cap \partial D_1 = \emptyset$. Fix $\delta_1 > 0$. Let $l \subset F_1$ be an arc joining the point a with the boundary ∂D_1 such that

$$\text{length}(l) < d(a, \partial D_1) + \delta_1.$$

But γ separates a and ∂D_1, and therefore, $\gamma \cap l \neq \emptyset$. Let $\xi \in \gamma \cap l$ be a point. We have

$$d(O_1) \quad \leq d(\xi, \partial D_1) + d(\xi, p) + d(p, O_1) \leq$$

$$\leq d(a, \partial D_1) + \delta_1 + \text{length}(\gamma) + d(p, O_1) \leq$$

$$\leq d(O_1) - \mu + d(p, O_1) + \delta_1.$$

From this we find

$$\mu \leq d(p, O_1) + \delta_1.$$

Because $\delta > 0$ is arbitrary, we obtain $\mu \leq d(p, O_1)$, and the necessary statement is proved.

So, every geodesic circle $S(p, \tau)$ separates the point p from O_1 for $\tau \in (\rho, \mu)$. Thus we are right to use Theorem 4.24 with $K = K(p, \rho, \mu)$. Every arc (or curve) $S_{D_1}(p, \tau)$ separates points a, b from ∂D_1 and O_1. Its image $hS_{D_1}(p, \tau)$ separates ha, hb from ∂D_2 and O_2. Therefore,

$$\rho(ha, hb; O_2, D_2) \leq \text{length}(hS_{D_1}(p, \tau)).$$

By (4.25) we find

$$\rho^2(ha, hb; O_2, D_2) \leq \left(\int\limits_{\rho}^{\mu} dt \bigg/ \int\limits_{S_D(p,t)} Q(y)\, |dy| \right)^{-1} \leq \omega(\rho);$$

and (4.33) is valid.

To complete the proof, we assume that $a_n, b_n \in D_1$ are points running to $a, b \in \tilde{D}_1 \setminus D_1$ and satisfying (4.32). We are right to assume that

$$\rho(ha_n, hb_n; O_2, D_2) \leq \omega^{1/2}(\rho(a_n, b_n; O_1, D_1)).$$

Now putting $n \to \infty$, we prove (4.33) in the general case. $\qquad \square$

Open questions 4.55 1) What is the relative distance on abstract surfaces with Finsler's metrics? 2) Study Q^*-mappings of abstract surfaces of the general form.

Chapter 5

The Speed of Approximation

Below in addition to Theorem 3.8, we give some estimates of the speed of an approximation of the canonic homeomorphism $F : B(0,R) \to \mathbb{R}^m$ with mappings F_n. To this end we use the concept of stability of conformal mappings in the class of mappings with the bounded Dirichlét integral. We use the statement saying that maps F_n and F are mutually close in the sense that they can be obtained one from another by small deformation. We give different estimates of such smallness.

5.1   Characteristics of a Closeness

Suppose that $p(x) : D \to [1, \infty)$ is a Lebesgue measurable function. As above in Theorem 3.8, we assume that a number sequence $\{Q_n\}$, $Q_n \geq 1$, $\lim\limits_{n \to \infty} Q_n \to \infty$ is fixed. Let

$$
\begin{aligned}
I_n \quad &= \{x \in D : p(x) \geq Q_n\}, \quad n = 1, 2, \dots, \\[1mm]
p_n(x) \quad &= 1 \quad \text{for} \quad x \in I_n, \\[1mm]
p_n(x) \quad &= p(x) \quad \text{for} \quad x \in D \setminus I_n.
\end{aligned}
$$

Our aim is the study of the convergence speed of homeomorphisms $w_n(x) :$ $D \to B(0, R_n)$, described in Section 3.1, with characteristics (p_n, θ) and the normalization (3.6) to the limiting homeomorphism $w(x) : D \to B(0, R)$ with

93

characteristics (p, θ). Below we will interpret this problem as a problem of the stability of a conformal mapping in the class BL, and under estimates of the stability order we will follow the articles [74], [193]. However, estimates of the stability order are not self-aim here. They supplement other quality characteristics of the approximating sequence $\{w_n(x)\}$. Therefore, our normalization for $w_n(x)$, $n = 1, 2, \ldots$, is different from normalizations considered traditionally, what inserts an additional specific.

Now let $\xi = w(x) : D \to B(0, R)$ be a limiting mapping. The inverse mapping $x = w^{-1}(\xi)$ with the composition of the mapping $\xi = w_n(x)$ is conformal on an open set $B = B(0, R) \setminus w(I_n)$ and quasi-conformal with characteristic which is not greater than $Q_n p(w^{-1}(\xi))$. We denote the superposition by the symbol

$$\zeta = h(\xi) = w_n \circ w^{-1}(\xi) : B(0, R) \to B(0, R_n), \quad \zeta = \eta_1 + i\,\eta_2.$$

In this place it is convenient to use complex destignations[1]

$$z = \xi_1 + i\,\xi_2, \quad \overline{z} = \xi_1 - i\,\xi_2.$$

If $\zeta = h(z)$ is a complex-valued function, then we consider it as the function $\zeta = h(z, \overline{z})$ and write

$$d\zeta = h'_z\, dz + h'_{\overline{z}}\, d\overline{z}$$

where by h'_z, $h'_{\overline{z}}$ we denote formal derivatives

$$h'_z = \frac{1}{2}\left(h'_{\xi_1} - i\,h'_{\xi_2}\right), \quad h'_{\overline{z}} = \frac{1}{2}\left(h'_{\xi_1} + i\,h'_{\xi_2}\right).$$

We recall the chain rule for the superposition $f \circ g$. We have

$$(f \circ g)_z = (f_w \circ g)\, g'_z + (g_{\overline{w}} \circ g)\, (\overline{g})'_z,$$

$$(f \circ g)_{\overline{z}} = (f_w \circ g)\, g'_{\overline{z}} + (g_{\overline{w}} \circ g)\, (\overline{g})'_{\overline{z}}. \tag{5.1}$$

Suppose that

$$\iint\limits_{B(0,R)} \left(|h_z|^2 + |h_{\overline{z}}|^2\right)\, d\xi_1 d\xi_2 < \infty.$$

As a closeness measure of a mapping

$$h = h_1 + i\,h_2 : B = B(0, R) \to B(0, R_n)$$

---

[1] See, for example, monographs [77], [130] where such complexification is developed very consistently.

to conformal mappings, we choose the quantity

$$\|h_{\overline{z}}\|_{L^2(B)} \equiv \left(\iint_B |h_{\overline{z}}|^2 \, d\xi_1 d\xi_2 \right)^{1/2} . \tag{5.2}$$

At points where h is differentiable, we have

$$|h_z|^2 = \frac{1}{4}|h_{\xi_1} - i\,h_{\xi_2}|^2 =$$

$$= \frac{1}{4} \left((h_{1\xi_1} + h_{2\xi_2})^2 + (h_{1\xi_2} - h_{2\xi_1})^2 \right) =$$

$$= \frac{1}{4} \left(h_{1\xi_1}^2 + h_{1\xi_2}^2 + h_{2\xi_1}^2 + h_{2\xi_2}^2 + 2J(z,h) \right) .$$

Here $J(z,h)$ is the Jacobian of h at a point z. Analogously,

$$|h_{\overline{z}}|^2 = \frac{1}{4}|h_{\xi_1} + i\,h_{\xi_2}|^2 =$$

$$= \frac{1}{4} \left((h_{1\xi_1} - h_{2\xi_2})^2 + (h_{1\xi_2} + h_{2\xi_1})^2 \right) =$$

$$= \frac{1}{4} \left(h_{1\xi_1}^2 + h_{1\xi_2}^2 + h_{2\xi_1}^2 + h_{2\xi_2}^2 - 2J(z,h) \right) .$$

Thus we have the following two useful formulas:

$$|h_z|^2 + |h_{\overline{z}}|^2 = \frac{1}{2} \left(h_{1\xi_1}^2 + h_{1\xi_2}^2 + h_{2\xi_1}^2 + h_{2\xi_2}^2 \right) , \tag{5.3}$$

and

$$|h_z|^2 - |h_{\overline{z}}|^2 = h_{1\xi_1} h_{2\xi_2} - h_{1\xi_2} h_{2\xi_1} = J(z,h) . \tag{5.4}$$

We calculate the characteristic $p_h = p_h(z)$ of the linear map $dh : \mathbb{R}^2 \to \mathbb{R}^2$. We observe at first that

$$dh = h_{\xi_1} d\xi_1 + h_{\xi_2} d\xi_2 =$$

$$= h_{\xi_1} \frac{dz + d\overline{z}}{2} + h_{\xi_2} \frac{dz - d\overline{z}}{2i} =$$

$$= \frac{1}{2} \left(h_{\xi_1} - i\,h_{\xi_2} \right) dz + \frac{1}{2} \left(h_{\xi_1} + i\,h_{\xi_2} \right) d\overline{z}.$$

Omitting simple calculations, we have

$$
\begin{aligned}
p_h(z) &= \frac{|h_z|^2 + |h_{\bar{z}}|^2}{|h_z|^2 - |h_{\bar{z}}|^2} + \sqrt{\left(\frac{|h_z|^2 + |h_{\bar{z}}|^2}{|h_z|^2 - |h_{\bar{z}}|^2}\right)^2 - 1} = \\[2mm]
&= \frac{|h_z|^2 + |h_{\bar{z}}|^2}{|h_z|^2 - |h_{\bar{z}}|^2} + \sqrt{\frac{\left(|h_z|^2 + |h_{\bar{z}}|^2\right)^2 - \left(|h_z|^2 - |h_{\bar{z}}|^2\right)^2}{\left(|h_z|^2 - |h_{\bar{z}}|^2\right)}} = \\[2mm]
&= \frac{|h_z|^2 + |h_{\bar{z}}|^2 + 2|h_z|\,|h_{\bar{z}}|}{|h_z|^2 - |h_{\bar{z}}|^2},
\end{aligned}
$$

that is

$$
p_h(z) = \frac{|h_z| + |h_{\bar{z}}|}{|h_z| - |h_{\bar{z}}|}. \tag{5.5}
$$

By the definition (5.2) of the closeness measure of h to conformal mappings and relations (5.3), (5.4), we obtain

$$
\|h_{\bar{z}}\|_{L^2(B)}^2 = \frac{1}{4} \iint\limits_{B(0,R)} \left(h_{1\xi_1}^2 + h_{1\xi_2}^2 + h_{2\xi_1}^2 + h_{2\xi_2}^2 - 2J(z,h)\right) d\xi_1 d\xi_2
$$

whence, using (5.3) and (5.4), we find

$$
\|h_{\bar{z}}\|_{L^2(B)}^2 = \frac{1}{2} \iint\limits_{B(0,R)} \left(\frac{|h_z|^2 + |h_{\bar{z}}|^2}{|h_z|^2 - |h_{\bar{z}}|^2} - 1\right) J(z,h)\, d\xi_1 d\xi_2 \le
$$

$$
\le \frac{1}{2} \iint\limits_{B(0,R)} (p_h(z) - 1)\, J(z,h)\, d\xi_1 d\xi_2 .
$$

Here we have used the estimate

$$
\frac{|h_z|^2 + |h_{\bar{z}}|^2}{|h_z|^2 - |h_{\bar{z}}|^2} \le \frac{(|h_z| + |h_{\bar{z}}|)^2}{|h_z|^2 - |h_{\bar{z}}|^2} = p_h(z). \tag{5.6}
$$

It is not difficult to see that since $h = w_n \circ w^{-1}$, then for all $z \in B(0,R) \setminus w(I_n)$ we have $p_h(z) = 1$, and for all $z \in w(I_n)$

$$
p_h(z) \le p_{w_n}(w^{-1}(z))\, p_{w^{-1}}(z).
$$

Therefore, for all $z \in w(I_n)$:

$$
p_h(w(x)) \le Q_n\, p_{w^{-1}}(w(x)) \le p_w^2(x).
$$

Thus, using $J(z, h) = J(x, w_n) J(z, w^{-1})$, we find

$$\|h_{\bar{z}}\|^2_{L^2(B)} \leq \frac{1}{2} \iint\limits_{w(I_n)} (p_w^2(z) - 1) \, J(x, w_n) J(z, w^{-1}) \, d\xi_1 d\xi_2 \,,$$

and further,

$$\|h_{\bar{z}}\|^2_{L^2(B)} \leq \frac{1}{2} \iint\limits_{I_n} (p^2(x) - 1) \, J(x, w_n) \, dx_1 dx_2 \,. \tag{5.7}$$

This estimate is the key for the following statement.

Lemma 5.8 *If for a number $n = 1, 2, \ldots$ the function*

$$P_n(x) = \begin{cases} p^2(x) - 1 & \text{for} \quad x \in I_n \,, \\ \\ 0 & \text{for} \quad x \in D \setminus I_n \end{cases}$$

has a majorant $K_n(x) \in W_0^{1,2}(D)$, then

$$\|h_{\bar{z}}\|^2_{L^2(B)} \leq 2 \, R_n^2 \iint\limits_{D} \left| \nabla K_n^{1/2}(x) \right|^2 p(x) \, dx_1 dx_2 \tag{5.9}$$

where

$$R_n = \sup_{x \in D} |w_n(x)| \,.$$

Proof is based on (5.7) and some constructions which are close to the proof of Lemma 3.19. We have

$$\iint\limits_{D} P_n \, J(x, w_n) \, dx_1 dx_2 \leq \iint\limits_{D} K_n \, J(x, w_n) \, dx_1 dx_2 \,.$$

Since K_n is finite on D, then by Green's formula we can write[2]

$$0 = \iint\limits_{\text{supp } K_n} K_n \, J(x, w_n) \, dx_1 dx_2 + \iint\limits_{\text{supp } K_n} \operatorname{Re} w_n \, dK_n \wedge d \operatorname{Im} w_n$$

---

[2] To verify these relations, it is sufficient to $W^{1,2}$-approximate K_n and w_n with smooth functions, to use Green's formula and to pass to the limit. We offer to make this standard procedure independently or to use corresponding reasonings from [98].

and

$$0 = \iint\limits_{\text{supp } K_n} K_n \, J(x, w_n) \, dx_1 dx_2 - \iint\limits_{\text{supp } K_n} \text{Im} \, w_n \, dK_n \wedge d\,\text{Re}\, w_n \,.$$

Here $w_n = \text{Re } w_n + i \, \text{Im } w_n = \eta_{n1} + i \, \eta_{n2}$.

From this we find

$$2 \iint\limits_{\text{supp } K_n} K_n \, J(x, w_n) \, dx_1 dx_2 \leq$$

$$\leq \iint\limits_{\text{supp } K_n} |\eta_{n1} \left(K_{nx_1} \eta_{n2x_2} - K_{nx_2} \eta_{2x_1} \right)| \; dx_1 dx_2 + \tag{5.10}$$

$$+ \iint\limits_{\text{supp } K_n} |\eta_{n2} \left(\eta_{n1x_1} K_{nx_2} - \eta_{n1x_2} K_{nx_1} \right)| \; dx_1 dx_2 \,.$$

Now we note that

$$|\eta_{n1} \left(K_{nx_1} \eta_{n2x_2} - K_{nx_2} \eta_{n2x_1} \right)| + |\eta_{n2} \left(\eta_{n1x_1} K_{nx_2} - \eta_{n1x_2} K_{nx_1} \right)| \leq$$

$$\leq |\eta_n| \left(\left(K_{nx_1} \eta_{n2x_2} - K_{nx_2} \eta_{n2x_1} \right)^2 + \left(\eta_{n1x_1} K_{nx_2} - \eta_{n1x_2} K_{nx_1} \right)^2 \right)^{1/2} \leq$$

$$\leq |w_n| \, |\nabla K_n| \left(|\nabla \eta_{n1}|^2 + |\nabla \eta_{n2}|^2 \right)^{1/2} \,.$$

By (5.3), (5.4), we have

$$|\nabla \eta_{n1}|^2 + |\nabla \eta_{n2}|^2 = 2 \frac{(|w_{nz}| + |w_{n\bar{z}}|)^2}{|w_{nz}|^2 - |w_{n\bar{z}}|^2} \, J(x, w_n) \,,$$

and using (5.6), we find

$$|\nabla \eta_{n1}|^2 + |\nabla \eta_{n2}|^2 \leq 2 \, p_n \, J(x, w_n) \,.$$

Thus from (5.10), it follows that

$$2 \iint\limits_{\text{supp}\,K_n} K_n\, J(x, w_n)\, dx_1 dx_2 \le$$

$$\le \iint\limits_{\text{supp}\,K_n} |w_n|\, |\nabla K_n|\, (2\, p_n\, J(x, w_n))^{1/2}\, dx_1 dx_2 \le$$

$$\le 2^{1/2} \left(\iint\limits_{\text{supp}\,K_n} |w_n|^2\, |\nabla K_n|^2\, \frac{p_n}{K_n}\, dx_1 dx_2 \right)^{1/2} \times$$

$$\times \left(\iint\limits_{\text{supp}\,K_n} K_n\, J(x, w_n)\, dx_1 dx_2 \right)^{1/2}.$$

From this we obtain

$$2 \iint\limits_{\text{supp}\,K_n} K_n\, J(x, w_n)\, dx_1 dx_2 \le \sup_D |w_n|^2 \iint\limits_{\text{supp}\,K_n} |\nabla K_n|^2\, \frac{p_n}{K_n}\, dx_1 dx_2\,,$$

and observing that $p_n(x) \le p(x)$ on D, we arrive to the inequality

$$\iint\limits_{\text{supp}\,K_n} K_n\, J(x, w_n)\, dx_1 dx_2 \le 2\, R_n^2 \iint\limits_{\text{supp}\,K_n} \left| \nabla K_n^{1/2} \right|^2 p(x)\, dx_1 dx_2\,.$$

The inequality, found before estimate (5.7), implies (5.9). The lemma is proved.
□

Remark 5.11 In the case if

$$\operatorname{ess\,sup}_{x \in D} p(x) < \infty$$

for sufficiently big n the sets $I_n = \emptyset$, we are right to choose $K_n \equiv 0$; and Lemma 5.8 is trivial.

□

The assumption $h \in W^1_{1,\text{loc}}$ and the equality $h_{\overline{z}} = 0$ imply holomorphy of h. The following stability problem is well-known. Suppose that h is topological and $h_{\overline{z}}$ is close to 0 in some sense. Is the map h close to a conformal mapping?

Because a conformal mapping of a disk onto itself with the normalization (3.6) is identical, then a closeness of h to a conformal mapping means, in this case, a closeness to the identical map. This means that, by small corrections (by a composition with h, or with h^{-1}), maps $w(\xi)$ and $w_n(\xi)$ can be transformed one by one, and the closeness measure of h to the identical map can be interpreted as a closeness measure between $w(\xi)$ and $w_n(\xi)$.

Assuming that the mappings $w(\xi)$ and $w_n(\xi)$ are automorphisms of disk $B(0, R)$, $R > 1$, normalized by conditions (3.6), we have the following theorem.

Theorem 5.12   *Suppose that the functions*

$$
P_n(x) = \begin{cases} p^2(x) - 1 & \text{for } x \in I_n, \\ \\ 0 & \text{for } x \in D \setminus I_n \end{cases}
$$

have majorants $K_n(x)$ of the class $W_0^{1,2}(B(0, R))$, $n = 1, 2, \ldots$.
   Let

$$
\delta_n = \left(\iint\limits_{B(0,R)} \left| \nabla K_n^{1/2}(x) \right|^2 p(x) \, dx_1 dx_2 \right)^{1/2}. \tag{5.13}
$$

There are $N \geq 1$ and $\varepsilon_0 > 0$ such that for all $n \geq N$ and $0 < \delta_n < \varepsilon_0$

$$
|h(x) - x| \leq C' \, \Delta(R\sqrt{2}\delta_n), \quad h = w_n \circ w^{-1}(x), \quad x \in B(0, R), \tag{5.14}
$$

and for every measurable set $E \subset B(0, R)$ the following estimate holds

$$
|\mathcal{H}^2(h(E)) - \mathcal{H}^2(E)| \leq C'' \, \Delta_1(R\sqrt{2}\delta_n). \tag{5.15}
$$

Here

$$
\Delta(\varepsilon) \equiv \frac{2 \, (2\pi \, k)^{\frac{1}{2}}}{\ln^{\frac{1}{2}} \varepsilon} + \frac{M}{2\sqrt{\pi}} \sqrt{\varepsilon}, \quad \varepsilon < \beta,
$$

$$
\Delta_1(\varepsilon) \equiv C \, \Delta(\varepsilon) + \sqrt{2\pi} R\varepsilon,
$$

$$
\beta = \left[\frac{1}{2} \min \left\{ 1, \frac{1}{4}R, \frac{1}{16}R^2 \right\} \right]^2,
$$

$$
k = \int\limits_{B(0,R)} |h'(x)|^2 dx_1 dx_2,
$$

$$
M = \text{const} \leq 8\pi^2 \, R.
$$

The proof of this theorem is based on stability theorems of conformal maps in the class of maps with bounded Dirichlét integral. The stability problem goes back to Lavrentiev [84] where the first qualitative solution was given for $(1+\varepsilon)$-quasi-conformal mappings of a disk onto a disk. The quantitative aspects of the Lavrentiev result were revised by Belinskii [13] — [15]. The further development of this problem is connected with papers of Krushkal [76], Suvorov [178], Lawrynowicz [87], Kruglikov, Miklyukov [74], Volynec [193] ($n = 2$) and with papers of Reshetnyak [148], Miklyukov [106], Belinski [16], Kopylov [70], [123] and others ($n > 2$).

In the proof of Theorem 3.1, we used the corresponding results of Kruglikov, Miklyukov [74] and Volynec [193].

5.2   Classes BL_k and BL

Let D' and D'' be subsets of $\mathbb{R}^2$. A topological mapping $w = f(z)$ of D' onto D'' belongs to a class BL_k if it has generalized Sobolev derivatives and

$$\iint\limits_{D'} \lambda^2(z, f)\, dx_1 dx_2 \leq k$$

where

$$\lambda(z, f) = \left[2(|f_z|^2 + |f_{\bar{z}}|^2)\right]^{1/2} .$$

A map f belongs to BL on D' if it belongs BL_k for some $0 \leq k < \infty$ (Suvorov [177, Chapter I]).

The functional class BL has been introduced by Beppo-Levy in 1906. On the history, see [177, §20].

We formulate in the necessary form some auxiliary statements on BL-mappings. Their proofs can be found in [177] or obtained independently as corollaries of general results of Chapter 4. We offer it to the reader as an exercise.

Below until the end of this chapter, we assume that the domains D' and D'' are simply connected and contain the origin.

Lemma 5.16   *Let $w = f(z)$ be a homeomorphic map of a class $BL_k(D')$, $z_0 \in D'$, and let $r < 2\min(1, |z_0, \partial D'|^2)$. Then for all $z \in \overline{B(z_0, r)}$ the following inequality holds*

$$|f(z) - f(z_0)| < (\pi k)^{\frac{1}{2}} \ln^{-\frac{1}{2}} \frac{2}{r} .$$

Here $|z, A|$ is the distance from a point z to a set A.

Lemma 5.17　*Let $w = f(z)$ be a homeomorphism of the class BL of a domain D' onto a domain D'', $f(0) = 0$ and $f^{-1} \in BL_k(D'')$. Let $F \subset D'$ be a continuum and $0 \in F$. Then for every pair of points $z_0 \in F$, $z \in D'$ such that $|z - z_0| \geq \alpha$, the following inequality holds*

$$|f(z) - f(z_0)| > 2 \exp \left\{ -\frac{\pi k}{\alpha^2} \right\} .$$

Here

$$\alpha = 2 \min \left\{ \frac{1}{2}d, \frac{3}{4}d^2, \exp \left\{ -\frac{\pi k}{[2 \min(1, \frac{1}{2}a, \frac{3}{4}a^2)]^2} \right\} \right\} ,$$

$$a = \min \left[\frac{1}{4}|0, \partial D''|^2, \exp \left(-\frac{4\pi k}{d^2} \right) \right] ,$$

$$d = |F, \partial D'|, \quad |A, B| - \text{distance between sets}\ \ A\ \text{and}\ B .$$

Lemma 5.18　*Let $w = f(z)$ be a BL_k-homeomorphism of a disk $B(0, R')$ onto a disk $B(0, R'')$, $1 < R', R'' < \infty$ and $f(0) = 0$, $f(1) = 1$. Then*

$$4 R'' \leq 4 + \frac{k}{\ln^{1/2} (R'/(R'-1))} .$$

Lemma 5.19　*Let $w = f(z)$ be a BL_k-homeomorphism of a disk $B(0, R')$ onto a disk $B(0, R'')$ and $f(0) = 0$. Then for every pair of points*

$$z_1, z_2 \in \overline{B(0, R')}, \quad |z_1 - z_2| \leq \frac{R'}{4} ,$$

it is fulfilled

$$|f(z_1) - f(z_2)| \leq 4R''(\pi k)^{\frac{1}{2}} \ln^{-\frac{1}{2}} \frac{R'}{|z_1 - z_2|} .$$

5.3　The Complex Dilatation

Let $\mu(z)$ be a Lebesgue measurable complex-valued function given a.e. on a domain D and such that $|\mu(z)| < 1$. Consider on D the Beltrami equation

$$f_{\overline{z}} = \mu(z) f_z \tag{5.20}$$

with the *complex dilatation* $\mu(z)$.

A continuous function $w = f(z)$ is called the BL_k-*solution* of (5.20) on D if $f \in BL_k(D)$ and satisfies (5.20) a.e. on D. A continuous function $w = f(z)$ is called the BL-*solution* if f is the BL_k-solution for some $k < \infty$.

Every schlicht BL-solution of (5.20) on D is a BL-homeomorphism of D into the $w = u + iv$-plane transforming infinitesimal ellipses with characteristics

$$p(z) = \frac{1 + |\mu(z)|}{1 - |\mu(z)|}, \quad \theta(z) = \frac{1}{2} \arg \mu(z)$$

into infinitesimal circles a.e. on D.

Conversely, every schlicht $BL(D)$-map $w = f(z)$, transforming infinitesimal ellipses with characteristics $p(z)$, $\theta(z)$ into infinitesimal circles, satisfies at points of differentiability the Beltrami equation (5.20) with the complex dilatation

$$\mu(z) = \frac{p(z) - 1}{p(z) + 1} e^{2i\theta(z)}.$$

The following statement is a special case of Theorem 2 of Kruglikov [72].

Theorem 5.21 *Let $w = f_n(z)$, $n = 1, 2, \ldots$, be a sequence of BL_k-solutions of (5.20) on a domain D with complex dilatations $\mu_n(z)$, respectively. Assume that the sequence $w = f_n(z)$ locally uniform on D converges to a BL_k-solution $w = f(z)$ of (5.20) with a complex dilatation $\mu(z)$. If*

$$\mu_n(z) \to \mu_\infty(z) \quad a.e. \ on \quad D,$$

then $\mu(z) = \mu_\infty(z)$ a.e. on D.

Proof. We use a scheme from the monograph [90, p. 187-189]. Without loss of generality we can assume that the limiting mapping is different from the identical constant and that the domains D and $f(D_n)$ uniformly bounded. It is sufficient to prove that

$$\zeta(z) = f_{\bar{z}}(z) - \mu_\infty(z) f_z(z) = 0 \quad a.e. \ in \quad D.$$

We have

$$\zeta = [f_{\bar{z}} - (f_n)_{\bar{z}}] + [(f_n)_{\bar{z}} - \mu_n(f_n)_z] +$$
$$+ [\mu_n(f_n)_z - \mu_\infty(f_n)_z] + [\mu_\infty(f_n)_z - \mu_\infty f_z].$$

Let R be a rectangle, $\overline{R} \subset D$, with sides parallel to coordinate axes, $R^* = R \cup D$. Then

$$\iint\limits_{R^*} \zeta(z)\,dxdy = I_{1,n} + I_{2,n} + I_{3,n} + I_{4,n}$$

where:

$$\begin{aligned}
I_{1,n} &= \iint_{R^*} [f_{\bar{z}} - (f_n)_{\bar{z}}]\,dxdy\,, \\
I_{2,n} &= \iint_{R^*} [(f_n)_{\bar{z}} - \mu_n(f_n)_z]\,dxdy\,, \\
I_{3,n} &= \iint_{R^*} [\mu_n(f_n)_z - \mu_\infty(f_n)_z]\,dxdy\,, \\
I_{4,n} &= \iint_{R^*} [\mu_\infty(f_n)_z - \mu_\infty f_z]\,dxdy\,.
\end{aligned}$$

We consider every integral separately.
(i_1)   Since

$$\iint\limits_{D} \lambda^2(z, f_n)\,dxdy \leq k\,, \quad n = 1, 2, \ldots\,,$$

by the theorem about weak convergence of generalized derivatives (see [171, p. 342]), we have

$$\lim_{n\to\infty} I_{1,n} = 0\,.$$

(i_2)   The integrals $I_{2,n}$ vanishes, because f_n are BL_k-solutions of (5.20) with complex dilatations μ_n, $n = 1, 2, \ldots$.

(i_3)   By the integral Cauchy inequality, we can write

$$|I_{3,n}|^2 \leq \iint\limits_{R^*} |(f_n)_z|^2\,dxdy \iint\limits_{R^*} |\mu_n - \mu_\infty|^2\,dxdy\,.$$

The first factor in the right side does not exceed k, and for the second factor, by Lebesgue convergence theorem we have

$$\lim_{n\to\infty} \iint\limits_{R^*} |\mu_n - \mu_\infty|^2\,dxdy = 0\,.$$

(i_4)   Now we remark that we can approximate $\mu_\infty(z)$ on D with step-functions $\alpha_i(z)$ $(i = 1, 2, \ldots)$, taking constant values on squares

$$R_{mq} = \{z = x + iy : (m-1)\,\delta < x < m\,\delta,\ (q-1)\,\delta < y < q\,\delta\}$$

where
$$m, q = 0, \pm 1, \pm 2, \ldots, \qquad \delta > 0, \ \delta = \delta(i) \to 0 \quad \text{as } i \to \infty$$

such that
$$\sup_z |\alpha_i(z)| \le \sup_z |\mu_\infty(z)|$$

and
$$\lim_{i \to \infty} \alpha_i(z) = \mu_\infty(z) \quad \text{a.e. in} \quad D.$$

We rewrite the integral $I_{4,n}$ in the following form

$$I_{4,n} = \iint_{R^*} (\mu_\infty - \alpha_i)\,[(f_n)_z - f_z]\,dxdy + \iint_{R^*} \alpha_i\,[(f_n)_z - f_z]\,dxdy.$$

For the first integral in the right side, we have

$$\left| \iint_{R^*} (\mu_\infty - \alpha_i)\,[(f_n)_z - f_z]\,dxdy \right|^2 \le$$

$$\le \iint_{R^*} |(f_n)_z - f_z|^2\,dxdy \iint_{R^*} |\mu_\infty - \alpha_i|^2\,dxdy$$

(see, for example, [90, p. 138]).

By Lebesgue theorem for an arbitrary $\varepsilon > 0$, there exists i_* such that

$$\iint_{R^*} |\mu_\infty - \alpha_{i_*}|^2\,dxdy \le \varepsilon.$$

In addition,

$$\iint_{R^*} |(f_n)_z - f_z|^2\,dxdy \le 2 \iint_{R^*} (|(f_n)_z|^2 + |f_z|^2)\,dxdy \le 4\,k.$$

On the other hand, as in the case (i_1) for every square $R_1 \subset R$ on which α_{i_*} is constant, the following equality holds

$$\lim_{n \to \infty} \iint_{R_1} \alpha_{i_*}[(f_n)_z - f_z]\,dxdy = 0.$$

Thus we proved that
$$\lim_{n \to \infty} I_{4,n} = 0$$

and

$$\iint\limits_{R} \zeta\, dxdy = 0. \tag{5.22}$$

It follows that ζ vanishes a.e. on D. To prove this, we first observe that each open set $O \subset D$ can be represented in the form

$$O = (\cup O_k) \cup S_0$$

where the sets O_k, $k = 1, 2, \ldots$, are disjoint horizontal rectangles and S_0 is a set with $\mathcal{H}^2(S_0) = 0$. By (5.22) the integral of ζ over O equals zero. If A, $\overline{A} \subset D$, is a measurable set, we can construct a decreasing sequence of open sets

$$O_k, \quad A \subset O_k, \quad \overline{O_k} \subset D, \quad k = 1, 2, \ldots,$$

such that $\mathcal{H}^2(O_k - A) = 0$. Since the Lebesgue integral is locally absolutely continuous as a set function, we conclude that

$$\iint\limits_{A} \zeta\, dxdy = 0 \tag{5.23}$$

for every measurable A, $\overline{A} \subset D$.

Denote by ξ the real part of ζ. Let

$$E = \{z : z \in D,\, \xi(z) \neq 0\}$$

be the union of the sets

$$E_k = \{z : z \in D,\, \xi(z) > 1/k\}, \quad E_{-k} = \{z : z \in D,\, \xi(z) < -1/k\}$$

where $k = 1, 2, \ldots$. Because every of these sets is of zero area by (5.23), we have $\mathcal{H}^2(E) = 0$, and so ξ vanishes a.e. in D. In the same way we show that the imaginary part of ζ vanishes a.e. in D, and the theorem is proved. $\qquad\square$

5.4   The Deviation on Compacts

Now let $w = f(z)$ be a topological mapping of the class BL of a domain D' onto D'' for which $f(0) = 0$ and

$$\|f_{\bar{z}}\|_{L^2(D')} \leq \varepsilon. \tag{5.24}$$

Lemma 5.25 *If for a mapping $w = f(z) : D' \to D''$ it is fulfilled (5.24) and the area $\mathcal{H}^2(D'') < \infty$, then $w = f(z)$ belongs to the class BL_k on D' with*

$$k = 2\,\mathcal{H}^2(D'') + 4\varepsilon^2\,.$$

For the **proof** it is sufficient to remark that

$$\iint\limits_{D'} \lambda^2(z,f)\,dx_1 dx_2 = \iint\limits_{D'} 2(|f_z|^2 + |f_{\bar z}|^2)\,dx_1 dx_2 =$$

$$= \iint\limits_{D'} 2(|f_z|^2 - |f_{\bar z}|^2)\,dx_1 dx_2 + 4\iint\limits_{D'} |f_{\bar z}|^2\,dx_1 dx_2\,.$$

By (5.4) we can write

$$\iint\limits_{D'} (|f_z|^2 - |f_{\bar z}|^2)\,dx_1 dx_2 = \iint\limits_{D'} J(z,f)\,dx_1 dx_2\,.$$

By the change of variables formula for BL-mappings, we have

$$\iint\limits_{D'} J(z,f)\,dx_1 dx_2 = \text{mes}_2\,(D'')\,,$$

and the lemma is proved. $\qquad\square$

Let F be a continuum in D' containing the origin. Let U_1, U_2 be subsets of D' such that $F \subset U_1$, $\overline{U_1} \subset U_2$, and having properties:

a) domains U_1, U_2 are simply connected;
b) boundaries ∂U_1, ∂U_2 are C^1-smooth;
c) $|F, \partial U_1| \geq \frac{1}{2} |F, \partial D'|$, $\quad |U_1, \partial U_2| \geq \frac{1}{4} |F, \partial D'|$.

There is the following integral representation for the mapping $w = f(z)$ in U_2

$$f(z) = h(z) - \frac{1}{\pi} \iint\limits_{U_2} \frac{f_{\bar z}\,d\eta_1 d\eta_2}{\zeta - z} \equiv h(z) + T_{U_2}(f_{\bar z}) \qquad (5.26)$$

where $h(z)$ is a holomorphic function in U_2 and $\zeta = \eta_1 + i\,\eta_2$ [189, Theorem **1.16**].

We will need following two properties of the operator $T_{U_2}(f)$ proofs of which can be found, for example, in [189, Chapter 1].

1. *If $f \in L^2(U_2)$, then*

$$\|T_{U_2}(f)\|_{L^2(U_2)} \leq M_{U_2}\,\|f\|_{L^2(U_2)} \qquad (5.27)$$

where

$$M_{U_2} = \sup_z \frac{1}{\pi} \left(\iint\limits_{U_2} \frac{d\eta_1 d\eta_2}{|\zeta - z|} \right)^2 .$$

2. *If $f \in L^2(U_2)$, then*

$$\left\| \frac{\partial T_{U_2}(f)}{\partial z} \right\|_{L^2(U_2)} \leq \|f\|_{L^2(U_2)} . \tag{5.28}$$

Remark 5.29 It is easy to see that

$$M_{U_2} = \sup_z \frac{1}{\pi} \left(\iint\limits_{U_2} \frac{d\eta_1 d\eta_2}{|\zeta - z|} \right)^2 \leq 4\pi \operatorname{diam}(U_2). \tag{5.30}$$

To check this inequality, it is sufficient to pass the integral to polar coordinates with the pole at the point z:

$$\iint\limits_{U_2} \frac{d\eta_1 d\eta_2}{|\zeta - z|} = \int\limits_0^{\operatorname{diam}(U_2)} d\tau \int\limits_{U_2 \cap |\zeta - z| = \tau} \frac{|dr|}{\tau} \leq 2\pi \operatorname{diam}(U_2).$$

$\square$

Exercise. Find an estimate for the constant M_U better than in the inequality (5.30). $\square$

Using the representation (5.26) for $w = f(z)$ and estimates (5.27), we have

$$\|f(z) - h(z)\|_{L^2(U_2)} \leq \|T_{U_2}(f_{\bar{z}})\|_{L^2(U_2)} \leq M_{U_2} \|f_{\bar{z}}\|_{L^2(U_2)} .$$

Hence, taking into account (5.24), we obtain

$$\|f(z) - h(z)\|_{L^2(U_2)} \leq M_{U_2} \varepsilon . \tag{5.31}$$

The relation (5.31) gives by itself an estimate of the deviation of f from a suitable holomorphic function h. Moreover, we need to have more information on this holomorphic function. We remark that

$$\iint\limits_{U_2} |h_z|^2\, dx_1 dx_2 \;=\; \iint\limits_{U_2} \left| f_z - \frac{\partial}{\partial z} T_{U_2}(f_{\bar z}) \right|^2 dx_1 dx_2 \le$$

$$\le 2 \iint\limits_{U_2} \left(|f_z|^2 + \left| \frac{\partial}{\partial z} T_{U_2}(f_{\bar z}) \right|^2 \right) dx_1 dx_2 \le \qquad (5.32)$$

$$\le 2 \iint\limits_{U_2} \left(|f_z|^2 + |f_{\bar z}|^2 \right) dx_1 dx_2 \le k$$

where k is the constant from Lemma 5.25.

Relations (5.31), (5.32) give a possibility to obtain an estimate for the difference $f(z) - h(z)$ with respect to the uniform metric in U_1. Fix a closed disk $\overline{B(z_0, r)}$ with the radius r and the center at $z_0 \in \overline{U_1}$. For $r < \frac{1}{4}|F, \partial D'|$, the disk $\overline{B(z_0, r)}$ is contained in the domain U_2. By the mean value theorem, there exists a point $z_1 \in \overline{B(z_0, r)}$ such that

$$|f(z_1) - h(z_1)|^2 \mathrm{mes}_2 \left(B(z_0, r) \right) = \|f(z) - h(z)\|^2_{B(z_0, r)} \le M^2_{U_2}\, \varepsilon^2 \,.$$

From this

$$|f(z_1) - h(z_1)| \le \frac{M_{D'}\, \varepsilon}{\sqrt{\pi}\, r} \,. \qquad (5.33)$$

At the point z_0 we have

$$|f(z_0) - h(z_0)| \le |f(z_0) - f(z_1)| + |h(z_0) - h(z_1)| + |f(z_1) - h(z_1)| \,.$$

By Lemma 5.16 and the inequality (5.33) for

$$r < \min \left\{ 1, \frac{1}{4}|F, \partial D'|, \frac{1}{16}|F, \partial D'|^2 \right\} \,,$$

we obtain

$$|f(z_0) - h(z_0)| < 2(\pi k)^{\frac{1}{2}} \ln^{-\frac{1}{2}} \frac{2}{r} + \frac{M_{D'}\, \varepsilon}{\sqrt{\pi}\, r} \,.$$

Now, setting

$$r = 2\sqrt{\varepsilon}$$

and taking into account arbitrariness under the choice of $z_0 \in \overline{U_1}$, we arrive to the inequality

$$|f(z) - h(z)| < \Delta(\varepsilon) \equiv \frac{2\,(2\pi\, k)^{\frac{1}{2}}}{\ln^{\frac{1}{2}} 1/\varepsilon} + \frac{M_{D'}}{2\sqrt{\pi}} \sqrt{\varepsilon}, \quad \varepsilon < \beta. \qquad (5.34)$$

Here

$$\beta = \left[\frac{1}{2} \min \left\{1, \frac{1}{4}|F, \partial D'|, \frac{1}{16}|F, \partial D'|^2\right\}\right]^2 .$$

Remark 5.35   Characters of ε, k and $M_{D'}$ are very different from each other and an estimate of the form

$$\Delta(\varepsilon) \leq \frac{\text{const}}{\ln^{\frac{1}{2}} 1/\varepsilon}$$

is not appropriate. If it is necessary, the reader can obtain a suitable estimate independently.   □

The estimate (5.34) gives the order of the deviation of the mapping $w = f(z)$ from a suitable holomorphic function $h(z)$. We show that there is $\varepsilon_0 > 0$ such that for all $0 < \varepsilon \leq \varepsilon_0$, the holomorphic function $h(z)$ in (5.34) is schlicht on F.

Suppose the contrary. Namely, we suppose that there exists a sequence of positive numbers $\{\varepsilon_n\}$, $\varepsilon_n \to 0$ as $n \to \infty$, such that the corresponding sequence of maps $\{f_n\}$ converges uniformly in D' to a limiting map f_0. By (5.34) we conclude that the corresponding sequence of holomorphic functions $\{h_n\}$ converges to f_0 uniformly on $\overline{U_1}$. Thus f_0 is holomorphic in U_1.

Using by Theorem 4.6 on the equicontinuous of the family BL_k in the domain closed with prime ends (see also Suvorov [177, Part I, §6, Theorem 9]), it is easy to verify that $f_0 \not\equiv$ const. Indeed, in the opposite case the normalization of $\{f_n\}$ implies $f_0 \equiv 0$ in D'. Now choose a sequence of points $\{z_m\}$, $m = 1, 2, \ldots$, in D', convergent to a prime end $e_0 \in \check{\partial}D'$. For all n we have

$$\lim_{m \to \infty} |f_n(z_m), \partial D''| = 0 .$$

For $n \to \infty$ this contradicts to the assumption $|0, \partial D''| > 0$.

Thus the holomorphic function f_0 is schlicht in $\overline{U_1}$ as an uniform limit of the sequence of homeomorphisms $\{f_n\}$. From this, in particular, it follows that $f_0^{-1} \in BL_{k_3}$ in the domain $\Delta = f_0(D')$ with $k_3 = 2\mathcal{H}^2(D')$.

By Lemma 5.17 for every pair of points $z_0 \in F$ and $z \in \partial U_1$, the following inequality holds

$$|f_0(z) - f_0(z_0)| > 2 \exp\left\{-\frac{\pi k_3}{\alpha^2}\right\}$$

where the constant $\alpha > 0$ is defined in Lemma 5.17. Observe that the sequence of holomorphic functions $\{h_n\}$ converges on $\overline{U_1}$ to f_0. Thus for every pair of points $z_0 \in F$ and $z \in \partial U_1$ for sufficiently large n, we have

$$|[h_n(z) - h_n(z_0)] - [f_0(z) - f_0(z_0)]| < 2 \exp\left\{-\frac{\pi k_3}{\alpha^2}\right\} <$$

$$< |f_0(z) - f_0(z_0)|.$$

By Rouché's theorem functions $h_n(z) - h_n(z_0)$ and $f_0(z) - f_0(z_0)$ have equal numbers of zeros in the domain U_1. Taking into account the arbitrariness of $z_0 \in F$, we conclude that the holomorphic functions $h_n(z)$ are schlicht for sufficiently big n. This contradicts to our assumption.

So, we showed that, for all $\varepsilon > 0$ lesser than some number $\varepsilon_0 > 0$, the function $h(z)$ in (5.34) is a schlicht conformal mapping on F.

Thus we proved the following statement.

Lemma 5.36 *Let $w = f(z)$ be a BL_k-homeomorphism of a domain D' onto a domain D'', $f(0) = 0$, $\|f_{\bar{z}}\|_{L^2(D')} \leq \varepsilon$ and $f^{-1} \in BL_{k_1}$ on D''. Then for every continuum F, $0 \in F \subset D'$, there exists a conformal mapping $h(z)$ such that, for all $0 < \varepsilon \leq \varepsilon_0$, the following inequality holds*

$$\max_{z \in F} |f(z) - h(z)| < \Delta(\varepsilon)$$

where ε_0 is a constant, depending on $|F, \partial D'|$, $|0, \partial D''|$, $\delta(D')$, k_1 and belonging to the segment $(0, \beta]$; the function $\Delta(\varepsilon)$ and the constant β are the same as in (5.34).

Now we can formulate the main result of Section 5.4.

Theorem 5.37 *Let D' and D'' be simply connected domains of the complex plane containing points 0 and 1. Let $w = f(z)$ be a BL_k-holomorphic mapping of D' onto D'' with the normalization $f(0) = 0$ and $f(1) = 1$. Assume that $\|f_{\bar{z}}\|_{L^2(D')} \leq \varepsilon$ and $f^{-1} \in BL_{k_1}(D'')$. Then for every continuum $F \subset D'$, $0, 1 \in F$, there exist $\varepsilon_0 > 0$ and a conformal mapping $h(z)$ with properties: $h(0) = 0$, $h(1) = 1$ and such that, for all $0 < \varepsilon \leq \varepsilon_0$, the following estimate holds*

$$\max_{z \in F} |f(z) - h(z)| < C_1 \Delta(\varepsilon)$$

where

$$C_1 = 2 (\operatorname{diam} D'' + 1),$$

and the function $\Delta(\varepsilon)$ is defined in (5.34).

Proof. By Lemma 5.36 for every continuum $F \subset D'$, $0, 1 \in F$, there exists a conformal mapping $h(z)$ such that for all $0 < \varepsilon \leq \varepsilon_0$, for each point $z \in F$, it is fulfilled

$$|f(z) - h(z)| < \Delta(\varepsilon).$$

We put

$$\tilde{h}(z) = h(z) - h(0) \quad \text{and} \quad \tilde{f}(z) = f(z) - f(0).$$

Define a function $g(z)$ by the ratio $g(z) = \tilde{h}(z)/\tilde{h}(1)$. It is clear that $g(z)$ is conformal on F and $g(0) = 0$, $g(1) = 1$.

The following two inequalities are simple corollaries of the showed estimate

$$|\tilde{f}(z) - \tilde{h}(z)| < 2\,\Delta(\varepsilon), \tag{5.38}$$

and

$$\left| \frac{\tilde{h}(z)}{\tilde{h}(1)} \right| < \frac{|\tilde{f}(z)| + 2\,\Delta(\varepsilon)}{|\tilde{f}(1)| - 2\,\Delta(\varepsilon)}. \tag{5.39}$$

The inequality (5.39) for

$$\Delta(\varepsilon) < \frac{1}{4}$$

gives

$$|\tilde{f}(1)| - 2\,\Delta(\varepsilon) = 1 - 2\,\Delta(\varepsilon) \geq \frac{1}{2},$$

and further,

$$\left| \frac{\tilde{h}(z)}{\tilde{h}(1)} \right| < 2\,\text{diam}\,D'' + 1. \tag{5.40}$$

Taking into account that $\tilde{f}(1) = 1$, we estimate the difference $f(z) - g(z)$. We have

$$|f(z) - g(z)| = \left| \frac{\tilde{h}(1)f(z) - \tilde{h}(z)}{\tilde{h}(1)} \right| \leq$$

$$\leq |\tilde{f}(z) - \tilde{h}(z)| + \left| \frac{\tilde{h}(z)}{\tilde{h}(1)} \right| |\tilde{h}(1) - \tilde{f}(1)|.$$

From this, using relations (5.38) and (5.40) for all $0 < \varepsilon \leq \varepsilon_0$, we obtain

$$\max_{z \in F} |f(z) - g(z)| < 2\,(\text{diam}\,D'' + 1)\,\Delta(\varepsilon).$$

Theorem is proved. □

Remark 5.41   The constant ε_0, as in Lemma 5.36, depends on $|F, \partial D'|$, $|0, \partial D''|$ and constants k, k_1. In [74], [193] there are some estimates of this constant. However, in our opinion, they are very rough. It is desirable to have more exact estimates.   □

5.5   Maps of a disk onto another disk

Our nearest aim is the proof of the following statement.

Theorem 5.42   *Let $w = f(z)$ be a BL_k-homeomorphism of a disk $B = B(0, R)$, $1 < R < \infty$, onto itself with the normalization $f(0) = 0, f(1) = 1$, and let $\|f_{\bar z}\|_{L^2(B(0,R))} \le \varepsilon$, $f^{-1} \in BL_{k_1}(B(0, R))$. Then there exists a constant $\varepsilon_0 > 0$ such that for all $0 < \varepsilon < \varepsilon_0$ and all $z \in B(0, R)$, the following estimate holds*

$$|f(z) - z| < C_2 \, \Delta(\varepsilon) \,. \tag{5.43}$$

Here C_2 is a constant depending on k, k_1, R and ε_0.

Proof. At first we remark that by Lemma 5.25 the map $w = f(z)$ belongs to BL_{k_1} in $B(0, R)$ with $k_1 = 2\pi R^2 + 4\varepsilon^2$. By Theorem 4.6 the map f is continuously extended up to a homeomorphism of closed disk $\overline{B(0, R)}$ onto itself.

We continue the mapping $w = f(z)$ by the symmetry with respect to circles $|z| = R$ and $|w| = R$. We put

$$f^*(z) = \begin{cases} R^2 \Big/ \overline{f}\left(\dfrac{R^2}{\bar z}\right) & \text{for} \quad |z| \ge R, \\[2mm] f(z) & \text{for} \quad |z| < R. \end{cases}$$

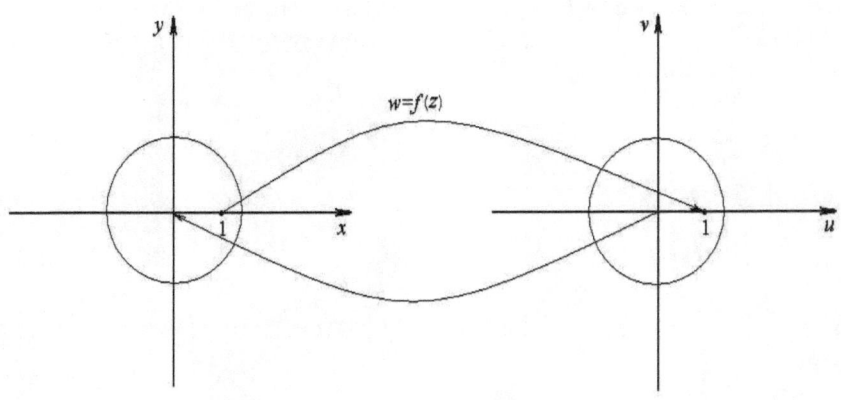

Fig. 5.1.

To calculate complex derivatives $(f^*)'_{\bar{z}}$ and $(f^*)'_{z}$, we use the chain rule for the superposition (5.1). Fix a constant ρ such that $R < \rho < \infty$. We have

$$\iint\limits_{R<|z|<\rho} |(f^*)'_{\bar{z}}(z)|^2 \, dx_1 dx_2 =$$

$$= \iint\limits_{R<|z|<\rho} \frac{R^4}{\left| \overline{f}^2 \left(\frac{R^2}{\bar{z}} \right) \right|^2} \left| (\overline{f})'_z \left(\frac{R^2}{\bar{z}} \right) \frac{R^2}{\bar{z}^2} \right|^2 \, dx_1 dx_2 =$$

$$= \iint\limits_{R<|z|<\rho} \frac{R^8}{\left| f \left(\frac{R^2}{\bar{z}} \right) \right|^4} \left| (\overline{f})'_z \left(\frac{R^2}{\bar{z}} \right) \right|^2 \frac{dx_1 dx_2}{|z|^4} ,$$

and since $(\overline{f})'_z = \overline{(f'_{\overline{z}})}$,

$$\iint\limits_{R<|z|<\rho} |(f^*)'_{\overline{z}}(z)|^2 \, dx_1 dx_2 \leq$$

$$\leq \iint\limits_{R<|z|<\rho} \frac{R^8}{\left|f\left(\dfrac{R^2}{\overline{z}}\right)\right|^4} \left|f'_{\overline{z}}\left(\dfrac{R^2}{\overline{z}}\right)\right|^2 \frac{dx_1 dx_2}{|z|^4} . \tag{5.44}$$

Fix a constant $\alpha > 1$. By Lemma 5.19 for all $z \in B(0, R)$ with

$$|z, \partial B(0, R)| \leq R/\alpha \,,$$

we have

$$|f(z), \partial B(0, R)| < 4R \, (\pi k)^{\frac{1}{2}} \, \ln^{-\frac{1}{2}} \frac{R}{|z, \partial B(0, R)|}$$

and

$$R - |f(z)| < 4R \, (\pi k)^{\frac{1}{2}} \, \ln^{-\frac{1}{2}} \frac{\alpha}{\alpha - 1} .$$

Thus for z satisfying the condition

$$R < |z| < \rho < \frac{\alpha}{\alpha - 1} R , \tag{5.45}$$

we find

$$R \geq \left|\frac{R^2}{\overline{z}}\right| \geq \frac{R^2}{\rho} \geq R\frac{\alpha - 1}{\alpha} .$$

From this we obtain

$$R \geq \left|f\left(\frac{R^2}{\overline{z}}\right)\right| \geq R\left(1 - 4\,(\pi k)^{\frac{1}{2}} \, \ln^{-\frac{1}{2}} \frac{\alpha}{\alpha - 1}\right) \equiv R\mu \tag{5.46}$$

where

$$\mu = 1 - 4\,(\pi k)^{\frac{1}{2}} \, \ln^{-\frac{1}{2}} \frac{\alpha}{\alpha - 1}$$

and the constant $\alpha \geq 4$ is chosen such that $\mu > 0$; for example,

$$\frac{\alpha}{\alpha - 1} > \exp\{16\,\pi\,k\} .$$

Relations (5.44), (5.45), (5.46) imply

$$\iint\limits_{R<|z|<\rho} |(f^*)'_{\bar z}(z)|^2 \, dx_1 dx_2 \le$$

$$\le \iint\limits_{R<|z|<\rho} \frac{R^4}{\mu^4} \left| f'_{\bar z}\left(\frac{R^2}{\bar z}\right)\right|^2 \frac{dx_1 dx_2}{|z|^4}.$$

By the substitution

$$\frac{R^2}{\bar z} = z^*,$$

we obtain

$$\iint\limits_{R<|z|<\rho} |(f^*)'_{\bar z}(z)|^2 \, dx_1 dx_2 \le$$

$$\le \frac{1}{\mu^4} \iint\limits_{\frac{R^2}{\rho}<|z|<R} |f'_{\bar z}(z)|^2 \, dx_1 dx_2. \tag{5.47}$$

Analogously,

$$\iint\limits_{R<|z|<\rho} |(f^*)'_z(z)|^2 \, dx_1 dx_2 =$$

$$= \iint\limits_{R<|z|<\rho} R^4 \left| \frac{1}{\bar f^2\left(\frac{R^2}{\bar z}\right)} \right| \left| (\bar f)'_{\bar z}\left(\frac{R^2}{\bar z}\right) \frac{R^2}{z^2}\right|^2 \, dx_1 dx_2 =$$

$$= \iint\limits_{R<|z|<\rho} \frac{R^8}{\left| f\left(\frac{R^2}{\bar z}\right)\right|^4} \left| (\bar f)'_{\bar z}\left(\frac{R^2}{z}\right)\right|^2 \frac{dx_1 dx_2}{|z|^4},$$

or by the relation $\overline{(f'_z)} = (\overline{f})'_{\overline{z}}$,

$$\iint\limits_{R<|z|<\rho} |(f^*)'_z(z)|^2\, dx_1 dx_2 \leq$$

$$\leq \iint\limits_{R<|z|<\rho} \frac{R^8}{\left|f\left(\frac{R^2}{\overline{z}}\right)\right|^4} \left|f'_z\left(\frac{R^2}{z}\right)\right|^2 \frac{dx_1 dx_2}{|z|^4}.$$

Reasoning as in the previous case and assuming that the conditions (5.45) are fulfilled with respect to R, ρ, α and $\mu > 0$, we have

$$\iint\limits_{R<|z|<\rho} |(f^*)'_z(z)|^2\, dx_1 dx_2 \leq \frac{1}{\mu^4} \iint\limits_{\frac{R^2}{\rho}<|z|<R} |f'_z(z)|^2\, dx_1 dx_2. \tag{5.48}$$

The estimate (5.47) guarantees that for the deviation measure of $f^*(z)$ from a holomorphic function in $B(0,\rho)$ it is fulfilled

$$\|(f^*)'_{\overline{z}}\|_{L^2(B(0,\rho))} \leq \left(\varepsilon^2 + \frac{1}{\mu^4} \iint\limits_{\frac{R^2}{\rho}<|z|<R} |f'_{\overline{z}}(z)|^2\, dx_1 dx_2\right)^{1/2}. \tag{5.49}$$

On the other hand, (5.47) and (5.48) imply that

$$\iint\limits_{R<|z|<\rho} \lambda^2(z,f^*)\, dx_1 dx_2 \leq \frac{2}{\mu^4} \iint\limits_{\frac{R^2}{\rho}<|z|<R} \lambda^2(z,f)\, dx_1 dx_2,$$

and in particular, the map $w = f^*(z)$ belongs to BL_{k_2} in $B(0,\rho)$ with a constant k_2 for which

$$k_2 \leq k + \frac{2k}{\mu^4}. \tag{5.50}$$

Fix ρ^* such that $R < \rho^* < \rho$. We will need some special estimate for the image of the circle $S_\rho = \{z : |z| = \rho^*\}$. By Theorem 5.37 for the continuum $\overline{B(0,\rho^*)}$, there exists a conformal map $w = h(z)$, $h(0) = 0$, $h(1) = 1$ such that for sufficiently small $\varepsilon > 0$, the following inequality holds

$$\max_{z \in \overline{B(0,\rho^*)}} |f(z) - h(z)| < C_1 \Delta(\varepsilon). \tag{5.51}$$

For $z \in S_R$ it is fulfilled $|f(z)| = R$. Thus for the holomorphic in $B(0, \rho^*)$ function $h(z)/z$ for $|z| = R$, we have

$$\frac{R - C_1 \Delta(\varepsilon)}{R} < \left| \frac{h(z)}{z} \right| = \frac{|h(z)|}{R} < \frac{R + C_1 \Delta(\varepsilon)}{R},$$

or

$$1 - \frac{C_1 \Delta(\varepsilon)}{R} < \left| \frac{h(z)}{z} \right| = \frac{|h(z)|}{R} < 1 + \frac{C_1 \Delta(\varepsilon)}{R}.$$

From here for $|z| = R$ and for sufficiently small $\varepsilon > 0$, we find

$$\frac{1}{1 + C_3 \Delta(\varepsilon)/R} < \left| \frac{h(z)}{z} \right| < 1 + C_3 \Delta(\varepsilon)/R$$

where

$$C_3 = C_3(\varepsilon) \equiv \frac{C_1}{1 - C_1 \Delta(\varepsilon)/R}.$$

The harmonic in $B(0, \rho^*)$ function

$$\ln \left| \frac{h(z)}{z} \right|$$

reaches its maximal and minimal values only on the boundary of a subdomain. Taking into account that the inequality $\ln(1 + \varepsilon) < \varepsilon$, we obtain

$$\left| \ln \left| \frac{h(z)}{z} \right| \right| < C_3 \Delta(\varepsilon)/R, \quad z \in \overline{B(0, R)}. \tag{5.52}$$

We use the relation

$$|\exp(\epsilon) - 1| \leq |\epsilon| \exp(|\epsilon|), \quad \epsilon \in \mathbb{R}. \tag{5.53}$$

Setting

$$\epsilon = \ln \left| \frac{h(z)}{z} \right|,$$

we find

$$\left| \left| \frac{h(z)}{z} \right| - 1 \right| \leq \left| \frac{h(z)}{z} \right| \left| \ln \left| \frac{h(z)}{z} \right| \right|.$$

By (5.52), we have

$$R \left| |h(z)| - |z| \right| < C_3 |h(z)| \Delta(\varepsilon), \quad z \in \overline{B(0, R)},$$

and for all $z \in \overline{B(0, R)}$,

$$\left| |h(z)| - |z| \right| < C_3 \left(1 + \frac{C_1 \Delta(\varepsilon)}{R} \right) \Delta(\varepsilon). \tag{5.54}$$

Estimates (5.51), (5.54) imply

$$\left| |f(z)| - |z| \right| < C_4 \, \Delta(\varepsilon), \quad C_4 = C_1 + C_3 \left(1 + \frac{C_1 \, \Delta(\varepsilon)}{R} \right),$$

or in other words, for all $z \in \overline{B(0, R)}$,

$$|z| - C_4 \, \Delta(\varepsilon) < |f(z)| < |z| + C_4 \, \Delta(\varepsilon).$$

Taking into account the construction of $f^*(z)$, we can now conclude that, for $|z| > R$, it is fulfilled

$$\frac{R^3}{R^3/|z| + C_4 \, \Delta(\varepsilon)} < |f^*(z)| < \frac{R^3}{R^3/|z| - C_4 \, \Delta(\varepsilon)}.$$

Thus, the image of the circle S_{ρ^*}, $R < \rho^* < \rho$, lies in the ring

$$\frac{\rho^*}{1 + C_4 \, \rho^* \, \Delta(\varepsilon)/R^3} < |w| < \frac{\rho^*}{1 - C_4 \, \rho^* \, \Delta(\varepsilon)/R^3}. \tag{5.55}$$

Enter upon, to the proof of the estimate (5.43). By Theorem 5.37, there exists a map $h^*(z)$ which is conformal in $B(0, \rho^*)$, normalized with conditions $h^*(0) = 0$, $h^*(1) = 1$ and such that, for all $z \in \overline{B(0, \rho^*)}$, it is fulfilled

$$|f^*(z) - h^*(z)| < C_1 \, \Delta(\varepsilon). \tag{5.56}$$

By (5.55) and (5.56), the curve $h^*(S_{\rho^*})$ lies in the ring

$$\frac{1}{1 + C_4 \rho^* \Delta/R^3} - \frac{C_1 \Delta}{\rho^*} < \left| \frac{w}{\rho^*} \right| < \frac{1}{1 - C_4 \rho^* \Delta/R^3} + \frac{C_1 \Delta}{\rho^*}$$

where $\Delta = \Delta(\varepsilon)$.

From here, for sufficiently small $\varepsilon > 0$, it follows that

$$\frac{1}{1 + C_5 \, \Delta(\varepsilon)} < \left| \frac{w}{\rho^*} \right| < 1 + C_5 \, \Delta(\varepsilon), \quad w \in f^*(S_{\rho^*}). \tag{5.57}$$

Exercise. Find the constant $C_5 = C_5(\varepsilon)$ and, further, the constants C_6, C_7. $\quad\square$

We put

$$w = P(\zeta) = h^* (\rho^* \, \zeta).$$

The map $\omega = P(\zeta)$ is conformal in the unit disk $|\zeta| < 1$. Moreover,

$$P(0) = 0, \quad P\left(\frac{1}{\rho^*} \right) = \frac{1}{\rho^*},$$

and by (5.57), the image of the circle $|\zeta| = 1$ lies in the ring

$$\frac{1}{1 + C_5 \, \Delta(\varepsilon)} < |w| < 1 + C_5 \, \Delta(\varepsilon) \,.$$

Since the function $\ln |P(\zeta)/\zeta|$ is harmonic in $|\zeta| < 1$ and for $|\zeta| = 1$

$$\frac{1}{1 + C_5 \, \Delta(\varepsilon)} < \left| \frac{P(\zeta)}{\zeta} \right| = |P(\zeta)| < 1 + C_5 \, \Delta(\varepsilon) \,,$$

then for all $|\zeta| \le 1$, we have

$$\left| \ln \left| \frac{P(\zeta)}{\zeta} \right| \right| < C_5 \, \Delta(\varepsilon) \,. \tag{5.58}$$

Consider the function

$$Q(\zeta) = \frac{1}{C_5 \, \Delta(\varepsilon)} \ln \frac{P(\zeta)}{\zeta}$$

where the symbol ln means the principal branch of the logarithm. The function $\tilde{Q}(\zeta) = Q(\zeta) - Q(0)$ is regular in the disk $|\zeta| < 1$, $\tilde{Q}(0) = 0$. By (5.58), we have

$$-1 < \mathrm{Re}\, \tilde{Q}(\zeta) < 1 \,.$$

Thus, by the Lindelöf principle [47, Chapter VIII, §3] for every $0 < r < 1$ and for all $|\zeta| \le r < 1$, we obtain

$$|\tilde{Q}(\zeta)| < \frac{2}{\pi} \ln \frac{1 + r}{1 - r} \,.$$

From here, for

$$r = \frac{R}{\rho^*} \,, \quad |\zeta| \le \frac{R}{\rho^*} \,,$$

we find

$$|\tilde{Q}(\zeta)| < \frac{2}{\pi} \ln \frac{\rho^* + R}{\rho^* - R} \,. \tag{5.59}$$

However, since

$$Q\left(\frac{1}{\rho^*} \right) = \frac{1}{C_5 \, \Delta(\varepsilon)} \ln \frac{P\left(1/\rho^* \right)}{1/\rho^*} = 0$$

and

$$Q(0) = Q\left(\frac{1}{\rho^*} \right) - \tilde{Q}\left(\frac{1}{\rho^*} \right) \,,$$

then we are right to conclude that

$$|Q(0)| \le \left| Q\left(\frac{1}{\rho^*} \right) \right| + \left| \tilde{Q}\left(\frac{1}{\rho^*} \right) \right| \le \frac{2}{\pi} \ln \frac{\rho^* + R}{\rho^* - R} \,. \tag{5.60}$$

Using again the inequality (5.53) for $x = |\ln P(\zeta)/\zeta|$, we have

$$|P(\zeta) - \zeta| \leq |\zeta| \left| \ln \frac{P(\zeta)}{\zeta} \right| \exp \left\{ \left| \ln \frac{P(\zeta)}{\zeta} \right| \right\} =$$

$$= C_5 \, \Delta(\varepsilon) \, |\zeta| \, |Q(\zeta)| \, \exp[\, C_5 \, \Delta(\varepsilon) \, |Q(\zeta)|] \leq$$

$$\leq C_5 \, \Delta(\varepsilon) \, |\zeta| \, [|\tilde{Q}(\zeta)| + |Q(0)|] \exp\{C_5 \, \Delta(\varepsilon) \, [|\tilde{Q}(\zeta)| + |Q(0)|]\}.$$

By Lemma 5.18 and (5.59), (5.60), from here, we deduce

$$|P(\zeta) - \zeta| < C_6 \, \Delta(\varepsilon).$$

Thus, for all $z \in B(0, R)$ and for sufficiently small $\varepsilon > 0$, we obtain

$$|h^*(z) - z| < C_7 \, \Delta(\varepsilon). \tag{5.61}$$

Now, the estimate (5.43) follows from (5.56) and (5.61). Theorem 5.42 is proved. □

Open questions 5.62   1) Find the best value of constants C_2 in the Theorem 5.42.   2) Find the best value of ε_0.

5.6   The Measure Stability

Theorem 5.42 implies series of corollaries which characterize other stability forms for BL-mappings (see [74]). We restrict ourselves to an estimate of measure distortion for maps close to conformal.

As above, let $w = f(z)$ be a BL-homeomorphism of a domain D. For every Lebesgue measurable set $E \subset D$, we have

$$\mathrm{mes}_2 f(E) = \iint\limits_{E} J(z, f) \, d\xi_1 d\xi_2 .$$

Using Theorem 5.42 and reasoning as in [106], we can obtain the following distortion estimate of the measure under BL-maps. For $(1+\varepsilon)$-quasi-conformal automorphisms of a disk and a different normalization, see Belinskii [14].

Theorem 5.63 *Let $w = f(z)$ be a BL_k-homeomorphism of a disk $B(0, R)$ onto itself, $1 < R < \infty$, with the normalization $f(0) = 0, f(1) = 1$, and let $\|f_{\bar{z}}\|_{L^2(B(0,R))} \leq \varepsilon$, $f^{-1} \in BL_{k_1}(B(0, R))$.*

Then, there exists a constant $\varepsilon_0 > 0$ such that, for all $0 < \varepsilon < \varepsilon_0$ and every measurable set $E \subset B(0, R)$, the following estimate holds

$$|\text{mes}_2\,(f(E)) - \text{mes}_2\,(E)| < C''\,\Delta_1(\varepsilon)\,. \tag{5.64}$$

Here, $\Delta_1(\varepsilon)$ is a function, defined in (5.65), and C'' is a constant, depending on k, k_1, R and ε_0.

Proof. We have

$$2\,\text{mes}_2\,(B(0, R)) = \int\limits_{S_R} \xi_1 d\xi_2 - \xi_2 d\xi_1 = \frac{1}{2i} \int\limits_{S_R} \bar{z}\,dz - z\,d\bar{z} =$$

$$= \frac{1}{2i} \int\limits_{S_R} \overline{[z - f(z)]}\,dz - \frac{1}{2i} \int\limits_{S_R} [z - f(z)]\,d\bar{z} +$$

$$+ \frac{1}{2i} \int\limits_{S_R} \overline{f(z)}\,dz - \frac{1}{2i} \int\limits_{S_R} f(z)\,d\bar{z}\,.$$

Employ the Green formula to two last terms

$$\frac{1}{2i} \int\limits_{\partial D} f(z)\,dz = \iint\limits_{D} f_{\bar{z}}\,d\xi_1 d\xi_2\,, \quad f \in C^0(\overline{D}) \cap BL(D)\,.$$

(The formula is valid for every finitely connected domain $D \subset \mathbb{R}^2$ with the smooth boundary.)

We have

$$2\,\text{mes}_2\,(B(0, R)) = \int\limits_{S_R} \xi_1 d\xi_2 - \xi_2 d\xi_1 = \frac{1}{2i} \int\limits_{S_R} \bar{z}\,dz - z\,d\bar{z} =$$

$$= \frac{1}{2i} \int\limits_{S_R} \overline{[z - f(z)]}\,dz - \frac{1}{2i} \int\limits_{S_R} [z - f(z)]\,d\bar{z} +$$

$$+ \iint\limits_{B(0,R)} (\bar{f})_{\bar{z}}\,d\xi_1 d\xi_2 + \iint\limits_{B(0,R)} f_{\bar{z}}\,d\xi_1 d\xi_2\,.$$

Thus,

$$2\,\mathrm{mes}_2\,(B(0,R)) \le \int\limits_{S_R} |z - f|\,|dz| + \iint\limits_{B(0,R)} \left(|(\overline{f})_{\overline{z}}| + |f_z| \right)\,d\xi_1 d\xi_2\,.$$

Evaluating the first term with Theorem 5.42 and transforming the second term with use the relation $|(\overline{f})_{\overline{z}}| = |f_z|$, we obtain

$$\pi\,R^2 \le C_8\,\Delta(\varepsilon) + \iint\limits_{B(0,R)} |f_z|\,d\xi_1 d\xi_2$$

where $C_8 = \pi\,C_2\,R$.

Let $E \subset B(0,R)$ be a Lebesgue measurable set. Then,

$$\mathrm{mes}_2\,(B(0,R)) - C_8\,\Delta(\varepsilon) \le \iint\limits_{E} |f_z|\,d\xi_1 d\xi_2 + \iint\limits_{B(0,R)\setminus E} |f_z|\,d\xi_1 d\xi_2\,.$$

Using the Cauchy integral inequality, we find

$$\mathrm{mes}_2\,(B(0,R)) - C_8\,\Delta(\varepsilon) \le$$

$$\le (\mathrm{mes}_2\,(E))^{\frac{1}{2}} \left(\iint\limits_{E} |f_z|^2\,d\xi_1 d\xi_2 \right)^{1/2} +$$

$$+ (\mathrm{mes}_2\,(B(0,R)\setminus E))^{\frac{1}{2}} \left(\iint\limits_{B(0,R)\setminus E} |f_z|^2\,d\xi_1 d\xi_2 \right)^{1/2}\,.$$

Now, we use the relation

$$\iint\limits_{E} |f_z|^2\,d\xi_1 d\xi_2 \le \iint\limits_{E} J(z,f)\,d\xi_1 d\xi_2 + \|f_{\overline{z}}\|^2_{L^2(B(0,R))}\,.$$

For $\|f_{\bar{z}}\|_{L^2(B(0,R))} \leq \varepsilon$, we have

$$\mathrm{mes}_2(B(0,R)) - C_8\Delta(\varepsilon) \leq (\mathrm{mes}_2(E))^{\frac{1}{2}} \left(\iint\limits_E J(z,f) d\xi_1 d\xi_2 + \varepsilon^2 \right)^{1/2} +$$

$$+ (\mathrm{mes}_2(B(0,R)) - \mathrm{mes}_2(E))^{\frac{1}{2}} \left(\iint\limits_{B(0,R)\backslash E} J(z,f) d\xi_1 d\xi_2 + \varepsilon^2 \right)^{1/2}.$$

However, by the inequality

$$\sqrt{a+b} \leq \sqrt{a} + \sqrt{b} \leq \sqrt{2(a+b)}, \quad a, b \geq 0,$$

it is fulfilled

$$(\mathrm{mes}_2(E))^{\frac{1}{2}} \left(\iint\limits_E J(z,f) d\xi_1 d\xi_2 + \varepsilon^2 \right)^{1/2} +$$

$$+ (\mathrm{mes}_2(B(0,R)) - \mathrm{mes}_2(E))^{\frac{1}{2}} \left(\iint\limits_{B(0,R)\backslash E} J(z,f) d\xi_1 d\xi_2 + \varepsilon^2 \right)^{1/2} \leq$$

$$\leq \varepsilon \sqrt{2\pi} R + (\mathrm{mes}_2(E))^{\frac{1}{2}} (\mathrm{mes}_2(f(E)))^{1/2} +$$

$$+ (\mathrm{mes}_2(B(0,R)) - \mathrm{mes}_2(E))^{\frac{1}{2}} (\mathrm{mes}_2(B(0,R)) - \mathrm{mes}_2(f(E)))^{1/2}.$$

Thus,

$$\mathrm{mes}_2(B(0,R)) - C_8\Delta(\varepsilon) \leq \varepsilon \sqrt{2\pi} R + (\mathrm{mes}_2(E))^{\frac{1}{2}} (\mathrm{mes}_2(f(E)))^{1/2} +$$

$$+ (\mathrm{mes}_2(B(0,R)) - \mathrm{mes}_2(E))^{\frac{1}{2}} (\mathrm{mes}_2(B(0,R)) - \mathrm{mes}_2(f(E)))^{1/2}.$$

Denoting $t = \mathrm{mes}_2(E)$ and $\omega = \mathrm{mes}_2(f(E)) - \mathrm{mes}_2(E)$, we can rewrite the last relation in the form

$$\pi R^2 - \Delta_1(\varepsilon) \leq t^{\frac{1}{2}} (\omega + t)^{\frac{1}{2}} + (\pi R^2 - t)^{\frac{1}{2}} (\pi R^2 - \omega - t)^{\frac{1}{2}}.$$

Here,

$$\Delta_1(\varepsilon) = C_8\,\Delta(\varepsilon) + \sqrt{2\pi}\,R\,\varepsilon\,,$$

or

$$\Delta_1(\varepsilon) = C_8\,\frac{2\,(2\pi\,k)^{\frac{1}{2}}}{\ln^{\frac{1}{2}}1/\varepsilon} + C_8\,\frac{M_{D'}}{2\sqrt{\pi}}\sqrt{\varepsilon} + \sqrt{2\pi}\,R\,\varepsilon\,. \tag{5.65}$$

Solving this inequality with respect to ω, for sufficiently small $\varepsilon > 0$, we obtain

$$|\omega| \le C''\,\Delta_1(\varepsilon)\,.$$

Theorem is proved. $\square$

5.7 Proof of Theorem 5.12

The arguments are not very complicated. If the quantity

$$\iint\limits_D \left|\nabla K_n^{1/2}(x)\right|^2 p(x)\,dx_1 dx_2$$

converges to zero as $n \to \infty$, then by Lemma 5.8 and Lemma 5.18, it can be used as the deviation of the mapping

$$h = w_n \circ w^{-1} : B(0,R) \to B(0,R)$$

to the identical mapping. Using (5.43) and (5.64), we arrive to necessary relations (5.14) and (5.15). $\square$

5.8 Remarks on $W^{1,2}$-Majorized Functions

In the present time, we have no explicit description of $W^{1,2}$-majorized functions. Here we notice only some their limit properties, arising from the theory of functions with generalized Sobolev derivatives.

5.8.1 The Set P_∞

From the definition of $W^{1,2}(D)$-functions, it follows that the *restrictions of $W^{1,2}(D)$-majorized functions P onto a.e. horizontal and vertical cuts of the domain D are locally bounded.*

Now we assume that the domain D is a disk and the function $P(x)$, defined by (3.9), is $W^{1,2}$-majorized in D. In other words, there exists a function $K(x) \in W^{1,2}(D)$ for which

$$P(x) \le K(x) \quad \text{for a.e.} \quad x \in D.$$

Denote by $L_1^2(\mathbb{R}^2)$ the set of the functions $\varphi(x)$ representable in the form of the Bessel potential

$$\varphi(x) = \iint\limits_{\mathbb{R}^2} G_1(|x - y|)\, u(y)\, dy_1 dy_2$$

where $u \in L^2(\mathbb{R}^2)$ is a function and $G_1(t)$ is the Bessel kernel of the order 1 (see the monograph of Adams and Hedberg [1, Section **1.2**]). Since D is a disk, then by the Calderon Theorem [1, Theorem **1.2.3**]) for every function $\varphi(x) \in W^{1,2}(D)$, there exists a function $\varphi^*(x) \in L_1^2(\mathbb{R}^2)$ such that a.e. in the disk $\varphi^*(x) = \varphi(x)$.

Thus since $K \in W^{1,2}(D)$, then there exists a function $u \in L^2(\mathbb{R}^2)$ such that

$$K(x) = \iint\limits_{\mathbb{R}^2} G_1(|x - y|)\, u(y)\, dy_1 dy_2 \quad \text{a.e. in the disk} \quad D.$$

By [1, Theorem **6.2.1**] the set

$$\{x \in D : \lim_{\zeta \to x} K(\zeta) = \infty\}$$

has the zero conformal capacity. Thus the following statement holds.

Theorem 5.66   *If P is $W^{1,2}$-majorized, then the set*

$$P_\infty = \{x \in D : \lim_{\zeta \to x} P(\zeta) = \infty\}$$

has the zero conformal capacity.

It is easy to see that this statement is valid for $W^{1,2}$-majorized functions P given over an arbitrary domain $D \subset \mathbb{R}^2$ different from the disk.

Thus the set P_∞ is sufficiently scarce for every $W^{1,2}$-majorized function $P : D \to \mathbb{R}$. In particular, the Hausdorff α-measure of P_∞ is equal to zero for an arbitrary $\alpha > 0$ (see [46, §**5.3**]).

5.8.2 The Class is not Empty

Now we show conditions to $p(x)$ for which the quantity $\delta_n \to 0$ in the domain $D = B(0, R)$. Thus the class of maps described by the Theorem 5.12 is not empty.

Theorem 5.67 *Let $p(x) : D \to [1, \infty]$ be a continuous (as a function acting from $D \subset \mathbb{R}^2$ to $\overline{R}$) function of the class $W^{1,2}(D)$.*

Then for the quantities δ_n defined in Theorem 5.12, it is fulfilled

$$\lim_{n \to \infty} \delta_n = 0 .$$

Proof. Let $(p(x), \theta(x))$ be characteristics in $D = B(0, R)$ and $p : D \to [1, +\infty]$ is continuous. Fix a number sequence $Q_n \to \infty$. As in Lemma 5.8, we define the sets I_n in D and, further, the functions P_n.

Since p is continuous, then the sets I_n are closed. Suppose that for the number $Q_n > 1$, the set $I_n \subset\subset D$. Without loss of generality, we can assume that $D \setminus I_n$ is a domain.

For every function K_n, admissible for calculations of δ_n, we have

$$\delta_n^2 \leq \iint\limits_D \left| \nabla K_n^{1/2}(x) \right|^2 p(x)\, dx_1 dx_2 =$$

$$= \iint\limits_{D \setminus I_n} \left| \nabla K_n^{1/2}(x) \right|^2 p(x)\, dx_1 dx_2 + \iint\limits_{I_n} |\nabla K_n(x)|^2\, \frac{p(x)}{K_n(x)}\, dx_1 dx_2 \leq$$

$$\leq Q_n \iint\limits_{D \setminus I_n} \left| \nabla K_n^{1/2}(x) \right|^2 dx_1 dx_2 + \iint\limits_{I_n} |\nabla K_n(x)|^2\, \frac{1 + K_n}{K_n}\, dx_1 dx_2 \leq$$

$$\leq Q_n \iint\limits_{D \setminus I_n} \left| \nabla K_n^{1/2}(x) \right|^2 dx_1 dx_2 + \left(1 + \frac{1}{Q_n} \right) \iint\limits_{I_n} |\nabla K_n(x)|^2\, dx_1 dx_2 .$$

Let $\varphi : D \setminus I_n \to \mathbb{R}$ be a function which is extremal under calculations of the capacity of the condenser $(I_n, \partial D; D)$, i.e. $\varphi = 0$ on ∂D, $\varphi = 1$ on I_n and

$$\iint\limits_{D \setminus I_n} |\nabla \varphi(x)|^2\, dx_1 dx_2 = \operatorname{cap}(I_n, \partial D; D) .$$

We choose K_n in the form

$$
K_n(x) = \begin{cases} Q_n\,\varphi^2(x) & \text{for} \quad x \in D \setminus I_n\,, \\[2mm] p(x) & \text{for} \quad x \in I_n\,. \end{cases}
$$

This function has the necessary properties and, as it is proved above,

$$
\delta_n^2 \le Q_n^2 \operatorname{cap}(I_n, \partial D; D) + \left(1 + \frac{1}{Q_n}\right) \iint\limits_{I_n} |\nabla p(x)|^2 \, dx_1 dx_2\,.
$$

Thus, if

$$
\lim_{n \to \infty} Q_n^2 \operatorname{cap}(I_n, \partial D; D) = 0\,, \tag{5.68}
$$

then $\delta_n \to 0$.

We show that the assumption $p(x) \in W^{1,2}(D)$ implies (5.68). Indeed, let

$$
A_{n_0,n} = \iint\limits_{I_{n_0} \setminus I_n} |\nabla p(x)|^2 \, dx_1 dx_2 < \infty\,.
$$

Suppose that the set $P_\infty = \{x \in D : p(x) = \infty\}$ is not empty. Choose $Q_{n_0} > 1$ such that the set I_{n_0} lies strongly inside D. Fix $Q_n > Q_{n_0}$. Consider the condenser $(D \setminus I_{n_0}, I_n; D)$. The function

$$
\varphi = \frac{1}{Q_n - Q_{n_0}} \, (p(x) - Q_{n_0})
$$

is admissible under the variational problem for the condenser capacity. Hence,

$$
\operatorname{cap}(I_n, \partial D; D) \le \operatorname{cap}(D \setminus I_{n_0}, I_n; D) \le \frac{A_{n_0,n}}{(Q_n - Q_{n_0})^2}\,.
$$

From this we obtain

$$
Q_n^2 \operatorname{cap}(I_n, \partial D; D) \le \frac{A_{n_0,n}\, Q_n^2}{(Q_n - Q_{n_0})^2}\,. \tag{5.69}
$$

Exercise. Find the direct proof of Theorem 5.66 based on (5.69).

We complete our reasonings. Setting $Q_{n_0} = \frac{1}{2} Q_n$, we verify that the property (5.68) holds. Thus $\delta_n \to 0$ as $n \to \infty$, and the theorem is proved. $\quad\square$

Open question 5.70 Describe the class of the continuous functions $p(x)$: $D \to [1, \infty]$ for which

$$
\lim_{n \to \infty} \delta_n = 0\,.
$$

Chapter 6

The Area Distortion

Below we bring some estimates of the area distortion under conformal mappings of two-dimensional surfaces in $\mathbb{R}^m$ onto the unit disk.

6.1 Graphs over Discs

Let Ω be a locally Lipschitz surface in $\mathbb{R}^m$. The Euclidean distance in $\mathbb{R}^m$ induces a distance on Ω and a measurement method of angles between smooth curves on Ω.

Suppose that there exists a conformal mapping $f : \Omega \to B$ of a surface Ω onto the unit disk

$$B = \{y = (y_1, y_2) : y_1^2 + y_2^2 < 1\}.$$

If a set $A \subset \Omega$ has the small measure $\mathrm{mes}_2(A)$, then the measure $\mathrm{mes}_2(T(A))$ is small. Below we try to describe the quantitative side of this appearance.

At first we assume that Ω is the graph of a Lipschitz function $\xi = f(x_1, x_2)$ defined over a simply connected and bounded subdomain D of the plane $\mathbb{R}^2$. Denote by j the orthogonal projection of Ω onto D. If f is differentiable at a point $(x_1, x_2) \in D$, then every infinitesimal circle on Ω with the center at $(x_1, x_2, \xi) \in \Omega$ is an infinitesimal ellipse in the plane $\mathbb{R}^2$ with the center at $j(x_1, x_2, \xi) = (x_1, x_2)$. Let $p(x_1, x_2)$ and $\theta(x_1, x_2)$ be characteristics of this ellipse. By the Rademacher theorem, the function f is differentiable a.e., characteristics are defined a.e. in D. Moreover, a.e. in the domain we have

$$p(x_1, x_2) = \sqrt{1 + |\nabla f(x_1, x_2)|^2}. \tag{6.1}$$

The surface Ω is simply connected and can be of the parabolic or hyperbolic conformal type. In other words, Ω can be conformally mapped onto the plane

$\mathbb{R}^2$ or onto the unit disk $B \subset \mathbb{R}^2$ respectively. Since the domain D is bounded, then the assumption

$$\operatorname*{ess\,sup}_{x \in D} |\nabla f(x)| < \infty \tag{6.2}$$

implies the hyperbolic type of Ω.

Suppose that f satisfies (6.2) and there exists a conformal mapping $T : \Omega \to B$. The map T induces a quasi-conformal mapping $y = \tau(x) : D \to B$ with characteristics $(p(x), \theta(x))$ a.e. in D and such that

$$T = j^{-1} \circ \tau^{-1} : B \to \Omega .$$

Moreover, a mapping τ is K-quasi-conformal in D if and only if

$$\operatorname*{ess\,sup}_{x \in D} |\nabla f(x)| \le \sqrt{K^2 - 1} .$$

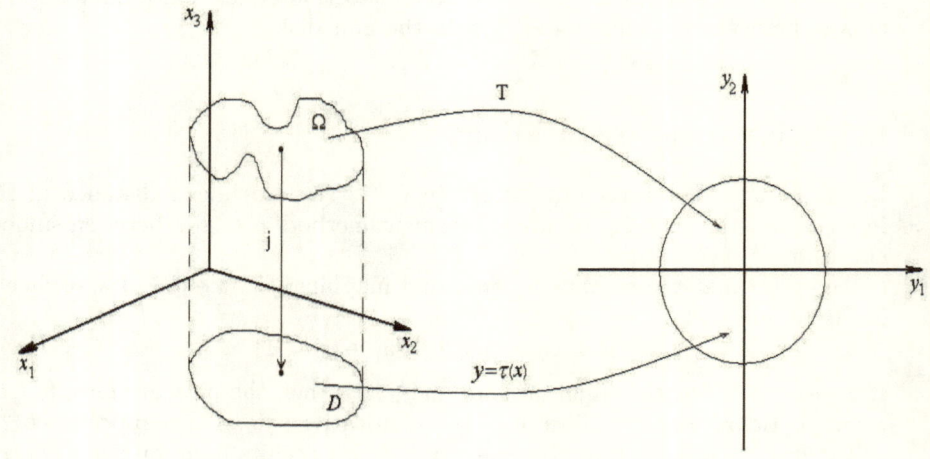

Fig. 6.1.

It is well-known (Astala [8]) that if D is the unit disk, then the assumption (6.2) implies that for every measurable set $A' \subset D$, it is fulfilled

$$\operatorname{mes}_2 (\tau(A')) \le C \, (\operatorname{mes}_2 (A'))^{1/K} \tag{6.3}$$

with a constant C depending only on K and the normalization $\tau^{-1}(0)$. This statement answers a question of Gehring and Reich [43]. The simplest proof of (6.3) is in [34].

For every measurable set $A \subset \Omega$, we have

$$\text{mes}_2\left(A\right) = \iint\limits_{A'} \sqrt{1 + |\nabla f|^2}\, dx_1 dx_2, \quad A' = j(A).$$

From this $\text{mes}_2\left(A'\right) \leq \text{mes}_2\left(A\right)$. Remarking that $T(A) = \tau(A')$ and using (6.3), we obtain

$$\text{mes}_2\left(T(A)\right) \leq C\left(\text{mes}_2\left(A\right)\right)^{1/K}. \tag{6.4}$$

The area estimate (6.4) is valid for each conformal map $T : \Omega \to B$ if:

(i) a surface $\Omega \subset \mathbb{R}^3$ is the graph of a function $\xi = f(x)$;

(ii) the function f is defined in a disk;

(iii) the gradient of f satisfies to (6.2).

If at least one of these assumptions (i) — (iii) violates, then such area estimate is not possible. Our aim is a study of the area distortion under conformal maps of surfaces $\Omega \subset \mathbb{R}^m$ of the general form.

6.2 K-Quasi-conformal Maps

Let Ω be a two-dimensional surface in $\mathbb{R}^m$, $2 \leq m < \infty$ given by a Lipschitz vector function (2.1) over the unit disk $B \subset \mathbb{R}^2$.

Suppose that the map $f : D \to \Omega$ is homeomorphic (with respect to the topology induced on Ω from $\mathbb{R}^m$). Moreover, at each point where f is differentiable, (3.1) holds.

The length element $|d\xi|$ in $\mathbb{R}^m$ induces the length element ds on the surface Ω such that

$$|d\xi|^2 = \sum_{i=1}^{m}\left(\sum_{j=1}^{2} f'_{ix_i}\, dx_j\right)^2 =$$

$$= g_{11}(\text{x})\, dx_1^2 + 2g_{12}(x)\, dx_1 dx_2 + g_{22}(x)\, dx_2^2$$

where

$$g_{11} = \sum_{i=1}^{m}\left(\frac{\partial f_i}{\partial x_1}\right)^2, \quad g_{12} = \sum_{i=1}^{m}\frac{\partial f_i}{\partial x_1}\frac{\partial f_i}{\partial x_2}, \quad g_{22} = \sum_{i=1}^{m}\left(\frac{\partial f_i}{\partial x_2}\right)^2.$$

The map $f : D \to \Omega$ is conformal if a.e. in D:

$$g_{11}(\text{x}) = g_{22}(\text{x}), \quad g_{12}(\text{x}) = 0. \tag{6.5}$$

Moreover, the coordinates x_1, x_2 are isothermal.

For every measurable set $A \subset \Omega$, we have

$$\text{mes}_2 (A) = \iint\limits_{A'} \sqrt{\text{g}_{11}\text{g}_{22} - \text{g}_{12}^2}\, dx_1 dx_2 \quad \text{where} \quad A' = j(A). \tag{6.6}$$

Suppose that the vector function (2.1) satisfies the following assumption at every point $x \in D$ of differentiability of f:

$$P(x) = \frac{\text{g}_{11}(\text{x}) + \text{g}_{22}(\text{x})}{2\sqrt{\text{g}_{11}(\text{x})\text{g}_{22}(\text{x}) - \text{g}_{12}^2(\text{x})}} \leq K < \infty \tag{6.7}$$

where K is a constant, $1 \leq K < \infty$.

By the Cauchy inequality and (6.6), for every measurable set $A' \subset D$, we have

$$\text{mes}_2 (A') = \iint\limits_{A'} dx_1 dx_2 \leq \left(\iint\limits_{A'} \sqrt{\text{g}_{11}\text{g}_{22} - \text{g}_{12}^2}\, dx_1 dx_2 \right)^{1/2} \times$$

$$\times \left(2 \iint\limits_{A'} \frac{\text{g}_{11} + \text{g}_{22}}{2\sqrt{\text{g}_{11}\text{g}_{22} - \text{g}_{12}^2}} \frac{dx_1 dx_2}{\text{g}_{11} + \text{g}_{22}} \right)^{1/2} \leq$$

$$\leq (2K)^{1/2} \left(\text{mes}_2 (A) \right)^{1/2} \left(\iint\limits_{A'} \frac{dx_1 dx_2}{\text{g}_{11} + \text{g}_{22}} \right)^{1/2}.$$

Using the cited Astala's result, we arrive to the following statement.

Theorem 6.8 *Let Ω be a surface in $\mathbb{R}^m$ given over the unit disk B with a Lipschitz vector function (2.1) satisfying (3.1), (6.7) and such that*

$$\iint\limits_{D} \frac{dx_1 dx_2}{\text{g}_{11} + \text{g}_{22}} \leq Q < \infty. \tag{6.9}$$

Let $T : \Omega \to B$ be a conformal mapping, $T(f(0)) = 0$.

For a measurable set $A \subset \Omega$, we have

$$\text{mes}_2 (T(A)) \leq C(K, Q) \left(\text{mes}_2 (A) \right)^{1/(2K)}. \tag{6.10}$$

Consider another example in which the constant $C(K,Q)$ is more precise. Suppose that Ω is the graph of a Lipschitz vector function of the form (2.17) defined in $\mathbb{R}^2$ and such that $f_3 = \ldots = f_m = 0$ outside of B.

It is easy to see that Ω has the parabolic type, i.e. for an arbitrary conformal map $T : \Omega \to \mathbb{R}^2$, it is fulfilled $T(\Omega) = \mathbb{R}^2$. This is clear, because the surface Ω is different from $\mathbb{R}^2$ only on a compact piece pasted along the unit circle.

In this case the conformal mapping $T : \Omega \to \mathbb{R}^2$ induces a quasi-conformal mapping $\tau(x) : \mathbb{R}^2 \to \mathbb{R}^2$. We find its coefficient of quasi-conformality K. We have

$$
g_{11} = 1 + \sum_{i=3}^{m} \left(\frac{\partial f_i}{\partial x_1} \right)^2, \quad g_{12} = \sum_{i=3}^{m} \frac{\partial f_i}{\partial x_1} \frac{\partial f_i}{\partial x_2}, \quad g_{22} = 1 + \sum_{i=3}^{m} \left(\frac{\partial f_i}{\partial x_2} \right)^2.
$$

Therefore,

$$
\frac{g_{11} + g_{22}}{\sqrt{g_{11}g_{22} - g_{12}^2}} \leq \frac{2 + \sum_{i=3}^{m} |\nabla f_i|^2}{\sqrt{1 + \sum_{i=3}^{m} |\nabla f_i|^2}},
$$

and we can change the condition (6.7) with the following requirement

$$
p(x) \leq \frac{1}{2} \left(\frac{1}{\sqrt{1 + \sum\limits_{i=3}^{m} |\nabla f_i(x)|^2}} + \sqrt{1 + \sum_{i=3}^{m} |\nabla f_i(x)|^2} \right) \leq K. \tag{6.11}
$$

Setting $j(x_1, x_2, \ldots, x_m) = (x_1, x_2)$, we have $T = j^{-1} \circ \tau^{-1} : \mathbb{R}^2 \to \Omega$. Since $f_3(x) = \ldots = f_m(x) = 0$ outside of B, then by (6.11) the characteristic $p(x) \equiv 1$ and τ is conformal outside of B. We assume that τ has the *hydrodynamic normalization*

$$
\tau(x) = x + O(1/|x|), \quad |x| \to \infty. \tag{6.12}
$$

The normalization (6.12) is very convenient, and there are many results of the conformal map theory, using this normalization.

Denote by

$$
U = \{x = (x_1, x_2) \in B : \nabla f_3(x_1, x_2) = \ldots = \nabla f_m(x_1, x_2) = 0\}
$$

the set of the critical points of the vector function $(f_3, \ldots, f_m) : \mathbb{R}^2 \to \mathbb{R}^{m-2}$ lying in B. For an arbitrary measurable set $A \subset U$, we have

$$
\text{mes}_2(A) = \iint\limits_{A'} dx_1 dx_2 \leq \pi
$$

where $A' = j(A)$ is the projection of A onto the (x_1, x_2)-plane.

By Theorem 1.6 of the article [9], the following statement holds.

Theorem 6.13 *Let Ω be a nonparametric Lipschitz surface given by a vector function $f : \mathbb{R}^2 \to \mathbb{R}^m$ of the form (2.17). Then for every measurable subset A of the set of the critical points U, it is fulfilled*

$$\left(\frac{1}{\pi} \mathrm{mes}_2\,(A) \right)^K \leq \frac{1}{\pi} \mathrm{mes}_2\,(T(A)) \leq \left(\frac{1}{\pi} \mathrm{mes}_2\,(A) \right)^{1/K} . \tag{6.14}$$

6.3   Main Theorem

Let Δ be a subset of a (y_1, y_2)-plane and let $g : \Delta \to \Omega$ be a conformal mapping. In isothermal coordinates y_1, y_2 the metric ds^2 of the surface Ω has the form

$$ds^2 = \Lambda(y)\,(dy_1^2 + dy_2^2) \quad \text{everywhere in} \quad \Delta$$

where $\Lambda(y) = E(y) = G(y)$.

If Ω is simply connected and has the hyperbolic type, then we will conclude that Δ is the unit disk $B = \{ y \in \mathbb{R}^2 : |y| < 1 \}$. The assumption (3.1) implies $\Lambda(y) > 0$ a.e. in Δ. By an auxiliary homothetic transformation in $\mathbb{R}^m$, we can assume that the map g satisfies the condition

$$\Lambda(0) = 1. \tag{6.15}$$

We denote by $K(\xi)$, the Gauss curvature of Ω at a point $\xi \in \Omega$.

The following statement supplies the main result of this chapter [122].

Theorem 6.16 *Let Ω be a surface in $\mathbb{R}^m$ of the nonnegative Gauss curvature, given over a domain $D \subset \mathbb{R}^2$ with a $W_{\mathrm{loc}}^{2,2}$-vector function of the form (2.1).*

Assume that f satisfies to (3.1) and

$$\iint\limits_{D} \frac{g_{11}(x) + g_{22}(x)}{\sqrt{g_{11}(x)g_{22}(x) - g_{12}^2(x)}}\, dx_1 dx_2 < \infty. \tag{6.17}$$

Then Ω has the hyperbolic conformal type, and for every conformal mapping $T : \Omega \to B$, satisfying (6.15) with $g = T^{-1}$ and for every measurable set $A \subset \Omega$, $\mathrm{mes}_2\,(A) < 1$, the following inequality holds

$$\mathrm{mes}_2\,(T(A)) \leq \frac{1 + \mathrm{mes}_2\,(\Omega)}{\ln 1\,/\mathrm{mes}_2\,(A)}. \tag{6.18}$$

6.4 Proof of Main Theorem

We use the idea of distortion estimates for boundary sets under conformal mappings a plane domain onto the disk (see [147, Chapter III, §1]). At first, we observe that the assumption (3.1) implies $df \neq 0$ at such point, the metric ds^2 is not degenerate, and an infinitesimal circle with the center at $x \in D$, given with respect to the metric ds^2 is an infinitesimal ellipse with respect to the Euclidean metric $|dx|^2 = dx_1^2 + dx_2^2$. Denote by (p, θ) characteristics of this ellipse, i.e. the ratio $p \geq 1$ of its axes and the angle θ, $0 \leq \theta < \pi$, between its largest axis and the direction $\overrightarrow{Ox_1}$. Moreover, it is easy to see that

$$p(x) = \frac{g_{11}(x) + g_{22}(x)}{2\sqrt{g_{11}(x)g_{22}(x) - g_{12}^2(x)}} + \left(\frac{(g_{11}(x) + g_{22}(x))^2}{4\left(g_{11}(x)g_{22}(x) - g_{12}^2(x)\right)} - 1 \right)^{1/2}.$$

Thus, there is a continuous field of characteristics (p, θ) in the domain D. By (3.1), we have

$$g_{11}(x)g_{22}(x) - g_{12}^2(x) > 0 \quad \text{everywhere} \quad D$$

and hence,

$$\sup_{D_1} p(x) < \infty \quad \text{for every subdomain} \quad D_1 \subset\subset D.$$

By the existence theorem, we find a quasi-conformal map $y = \tau(x) \to \mathbb{R}^2$ with characteristics (p, θ) a.e. in D. This mapping is unique up to conformal automorphisms $y' = \phi(y)$ on the (y_1, y_2)-plane.

In the general case, the image $\tau(D)$ can be the plane $\mathbb{R}^2$ or its proper subdomain. The assumption (6.17) implies that always

$$\iint_D p(x)\, dx_1 dx_2 = \iint_{\tau(D)} p(\tau^{-1}(y)) J(y, \tau^{-1})\, dy_1 dy_2 \leq$$

$$\leq \iint_{\tau(D)} |\nabla \tau^{-1}(y)|^2\, d\xi_1 d\xi_2 < \infty$$

where $J(y, \tau^{-1})$ is the Jacobian of τ^{-1} at $y \in \tau(D)$.

Consider a ring

$$A(t', t'') = \{y \in \mathbb{R}^2 : 0 < t' < |y| < t'' < \infty\}.$$

Now, we use (once again!) the Length and Area Principle (see Theorem 1.51 for the special choice of $H = \sigma \equiv 1$).

Lemma 6.19 *Let $g : A(t', t'') \to \mathbb{R}^2$ be a vector function of the class $W^{1,2}(A(t', t'))$. Then,*

$$\int_{t'}^{t''} \mathrm{osc}^2\, (g, S(0, r))\, \frac{dr}{r} \leq c_1\, I(g; t', t'') \qquad (6.20)$$

where c_1 is an absolute constant, and

$$I(g; t', t'') = \iint_{A(t', t'')} |\nabla g|^2\, dy_1 dy_2.$$

Suppose that $\tau(D) = \mathbb{R}^2$. We choose a ring $A(t', t'')$ and use the Length and Area Principle (6.20). We have

$$\inf_{r \in (t', t'')} \mathrm{osc}\, (\tau^{-1}, S(0, r)) \leq c_1^{1/2} I^{1/2}(\tau^{-1}; t', t'') \ln^{-1/2} \frac{t''}{t'}.$$

The map τ^{-1} is homeomorphic, and hence,

$$\mathrm{osc}\, (\tau^{-1}, B(0, r)) \leq \mathrm{osc}\, (\tau^{-1}, S(0, r)) \quad (0 < r < \infty).$$

From this,

$$\inf_{r \in (t', t'')} \mathrm{osc}\, (\tau^{-1}, B(0, r)) \leq c_1^{1/2} I^{1/2}(\tau^{-1}; t', t'') \ln^{-1/2} \frac{t''}{t'},$$

and setting $t'' \to \infty$, we obtain

$$\mathrm{osc}\, (\tau^{-1}, B(0, r)) = 0 \quad \text{for all} \quad 0 < r < \infty,$$

i.e., $\tau^{-1} \equiv \mathrm{const}$. We arrive to the contradiction.

Thus, the domain $\tau(D)$ is a proper simply connected subdomain of $\mathbb{R}^2$, and Ω is of the hyperbolic type.

Let $T : \Omega \to B$ be a conformal mapping. By (6.5), the length element on the surface Ω in isothermal coordinates y_1, y_2 has the form

$$ds^2 = \Lambda(y)\, (dy_1^2 + dy_2^2)$$

where $\Lambda = E = G > 0$ is a function.

Remark that since $f \in C^3(D)$, then $(p, \theta) \in C^2(D)$ and the map $\tau \in C^3(D)$. Here, $\Lambda \in C^2(B)$.

Let $E \subset \Omega$ be a measurable set. The set $e = T(A) \subset B$ is also measurable and

$$\mathrm{mes}_2\, (A) = \iint_e \Lambda(y)\, dy_1 dy_2 \leq |\Omega|.$$

Here and below, for brevity, let $\text{mes}_2\,(\Omega) = |\Omega|$. For an arbitrary $a > 0$, we put

$$\ln^+ a = \max\{\ln a, 0\}, \quad \ln^- a = \max\{-\ln a, 0\}.$$

It is clear that

$$\ln a = \ln^+ a - \ln^- a.$$

Since $\ln^+ a < a$, then we are right to assume that

$$|\Omega| \geq \int_B \Lambda(y)\,dy_1 dy_2 \geq \int_B \ln^+ \Lambda(y)\,dy_1 dy_2. \tag{6.21}$$

On the other hand,

$$\iint_B \ln \Lambda(y)\,dy_1 dy_2 = \iint_B \ln^+ \Lambda(y)\,dy_1 dy_2 - \iint_B \ln^- \Lambda(y)\,dy_1 dy_2$$

and, further, by (6.21), we obtain

$$-\iint_B \ln^- \Lambda(y)\,dy_1 dy_2 = \iint_B \ln \Lambda(y)\,dy_1 dy_2 - \iint_B \ln^+ \Lambda(y)\,dy_1 dy_2 \geq$$

$$\geq \iint_B \ln \Lambda(y)\,dy_1 dy_2 - |\Omega|\,.$$

$$\tag{6.22}$$

Using the well-known inequality

$$\frac{1}{\text{mes}_2\,(e)} \iint_e \varphi(y)\,dy_1 dy_2 \geq \exp\left\{\frac{1}{\text{mes}_2\,(e)} \iint_e \ln \varphi(y)\,dy_1 dy_2\right\},$$

we have

$$\text{mes}_2\,(A) = \iint_e \Lambda(y)\,dy_1 dy_2 =$$

$$= \text{mes}_2\,(e) \left(\frac{1}{\text{mes}_2\,(e)} \iint_e \Lambda(\xi)\,d\xi_1 d\xi_2\right) \geq$$

$$\geq \text{mes}_2\,(e)\exp\left\{\frac{1}{\text{mes}_2\,(e)} \iint_e \ln \Lambda(y)\,dy_1 dy_2\right\}.$$

However,

$$\iint\limits_{e} \ln \Lambda(y)\, dy_1 dy_2 \;=\; \iint\limits_{e} \ln^{+} \Lambda(y)\, dy_1 dy_2 - \iint\limits_{e} \ln^{-} \Lambda(y)\, dy_1 dy_2 \geq$$

$$\geq - \iint\limits_{e} \ln^{-} \Lambda(y)\, dy_1 dy_2 \geq$$

$$\geq - \iint\limits_{B} \ln^{-} \Lambda(y)\, dy_1 dy_2 \,,$$

and we arrive to the inequality

$$\mathrm{mes}_2\,(A) \geq \mathrm{mes}_2\,(e) \exp\left\{ -\frac{1}{\mathrm{mes}_2\,(e)} \iint\limits_{B} \ln^{-} \Lambda(y)\, dy_1 dy_2 \right\}.$$

Now the estimate (6.22) implies

$$\mathrm{mes}_2\,(A) \geq \mathrm{mes}_2\,(e) \exp\left\{ \frac{\iint\limits_{B} \ln \Lambda(y)\, dy_1 dy_2 - |\Omega|}{\mathrm{mes}_2\,(e)} \right\}. \tag{6.23}$$

Calculate the integral. By the assumption (3.1), we find $\Lambda > 0$ in B. Observe that

$$\frac{d}{dr} \int\limits_{S(0,r)} \ln \Lambda(y)\, |dy| \;=\; \frac{d}{dr} \int\limits_{0}^{2\pi} \ln \Lambda(r,\theta)\, r\, d\theta =$$

$$= \int\limits_{0}^{2\pi} r\frac{d}{dr} \ln \Lambda(r,\theta)\, d\theta + \int\limits_{0}^{2\pi} \ln \Lambda(r,\theta)\, d\theta =$$

$$= \int\limits_{S(0,r)} \frac{\partial}{\partial \mathbf{n}} \ln \Lambda(y)\, |dy| + \frac{1}{r} \int\limits_{S(0,r)} \ln \Lambda(y)\, |dy|$$

where $\frac{\partial}{\partial \mathbf{n}}$ means the normal derivative.

Thus by Green's formula we have

$$\frac{d}{dr} \int\limits_{S(0,r)} \ln \Lambda(y) \, |dy| \;\; = \iint\limits_{B(0,r)} \Delta \ln \Lambda(y) \, dy_1 dy_2 +$$

$$+ \frac{1}{r} \int\limits_{S(0,r)} \ln \Lambda(y) \, |dy| \, . \tag{6.24}$$

Analogously,

$$\lim_{r \to 1} \frac{1}{r} \int\limits_{S(0,r)} \ln \Lambda(y) \, |dy| - \lim_{r \to 0} \frac{1}{r} \int\limits_{S(0,r)} \ln \Lambda(y) \, |dy| =$$

$$= \int\limits_0^1 dr \left(\int\limits_0^{2\pi} \ln \Lambda(r,\theta) \, d\theta \right)' = \int\limits_0^1 dr \int\limits_0^{2\pi} \frac{d}{dr} \ln \Lambda(r,\theta) \, d\theta =$$

$$= \int\limits_0^1 \frac{dr}{r} \int\limits_{S(0,r)} \frac{\partial}{\partial \overline{n}} \ln \Lambda(y) \, |dy| = \int\limits_0^1 \frac{dr}{r} \iint\limits_{B(0,r)} \Delta \ln \Lambda(y) \, dy_1 dy_2 \, ,$$

and further,

$$\lim_{r \to 1} \int\limits_{S(0,r)} \ln \Lambda(y) \, |dy| \;\; = 2\pi \ln \Lambda(0) +$$

$$+ \int\limits_0^1 \frac{dr}{r} \iint\limits_{B(0,r)} \Delta \ln \Lambda(y) \, dy_1 dy_2 \, . \tag{6.25}$$

From this, using (6.24) and integrating by parts, we arrive to the relation

$$\iint_B \ln \Lambda(y)\, dy_1 dy_2 = \int_0^1 dr \int_{S(0,r)} \ln \Lambda(y)\, |dy| =$$

$$= \lim_{r \to 1} \int_{S(0,r)} \ln \Lambda(y)\, |dy| - \int_0^1 r\, dr \frac{d}{dr} \int_{S(0,r)} \ln \Lambda(y)\, |dy| =$$

$$= \lim_{r \to 1} \int_{S(0,r)} \ln \Lambda(y)\, |dy| - \int_0^1 r\, dr \iint_{B(0,r)} \Delta \ln \Lambda(y)\, dy_1 dy_2 -$$

$$- \int_0^1 dr \int_{S(0,r)} \ln \Lambda(y)\, |dy|\, .$$

Hence,

$$\iint_B \ln \Lambda(y)\, dy_1 dy_2 \;= \tfrac{1}{2} \lim_{r \to 1} \int_{S(0,r)} \ln \Lambda(y)\, |dy| -$$

$$-\frac{1}{2} \int_0^1 r\, dr \iint_{B(0,r)} \Delta \ln \Lambda(y)\, dy_1 dy_2\, .$$

Combining these relations with (6.25), we obtain

$$\iint_B \ln \Lambda(y)\, dy_1 dy_2 \;= \pi \ln \Lambda(0) +$$

$$+\frac{1}{2} \int_0^1 (\frac{1}{r} - r)\, dr \iint_{B(0,r)} \Delta \ln \Lambda(y)\, dy_1 dy_2\, . \tag{6.26}$$

By the Gauss theorem, the curvature K of Ω can be calculated with coefficients E, F, G and their partial derivatives up to the second order. In isothermal coordinates, it has the simplest form

$$K = -\frac{1}{2\Lambda} \Delta \ln \Lambda.$$

By (6.15) and (6.26), we have

$$\iint\limits_{B} \ln \Lambda(y)\, dy_1 dy_2 \geq 0.$$

Thus, the relation (6.23) can be rewritten as

$$\mathrm{mes}_2\,(A) > \mathrm{mes}_2\,(e) \exp\left\{-\frac{|\Omega|}{\mathrm{mes}_2\,(e)}\right\}. \tag{6.27}$$

Let $\mathrm{mes}_2\,(A) < 1$. By (6.27), we find

$$\ln \frac{1}{\mathrm{mes}_2\,(A)} \leq \frac{|\Omega|}{\mathrm{mes}_2\,(e)} + \ln \frac{1}{\mathrm{mes}_2\,(e)}.$$

Thus, we obtain finally

$$\mathrm{mes}_2\,(e) \leq \frac{|\Omega| + 1}{\ln \dfrac{1}{\mathrm{mes}_2\,(A)}}$$

that is equivalent to (6.18). Theorem is proved. □

Open questions 6.28   1) Prove Theorem 6.16 for non-regular surfaces of more general forms.   2) Find geometric conditions which guarantee the condition (6.15).

Chapter 7

Ahlfors-Warschawski Theorems

We introduce the concept of a reduced modulus of a simply connected domain with respect a 'boundary point' and give its applications in estimates close to the boundary of conformal mappings. In particular, we prove Ahlfors and Warschawski type theorems for conformal maps of plane bands.

Close boundary estimates of conformal maps with respect to the Euclidean metric, see [2], [195], [196], [197], [107], [108], [153]-[158] and [146, Chapter 11].

7.1 Plane Strips

Sometimes, it is not obligatory to know exactly the function which conformally maps a domain onto the unit ball or another canonical domain. Often, it is sufficient to have a behavior estimate of the function close to boundary points. Below, we recall two such classical results for conformal maps of plane domains [35, Chapter V, §6].

Let D be a simply connected subdomain of the $x = (x_1, x_2)$-plane. We will assume until this section that D has nonempty intersection with every vertical line $x_1 = \alpha$, $-\infty < \alpha < \infty$.

Let $y(x) = (y_1(x), y_2(x))$ be a function conformally mapped D onto a strip $|y_2| < \frac{\pi}{2}$ such that $y_1(x) \to \pm\infty$ for $x_1 \to \pm\infty$. By $x(y)$, we denote the function inverse to $y(x)$. It is clear that the function $y(x)$ is defined up to a real constant.

The boundary of D has two components: the upper C^+ and lower C^-. Under $y = y(x)$, the arc C^+ is mapped onto the straight line $y_2 = \frac{\pi}{2}$, and the arc C^- onto $y_2 = -\frac{\pi}{2}$.

Consider a cross-section of D with a vertical line $x_1 = \alpha$. In the general case, the cross-section contains a countable number of segments. We choose segments which join C^+ and C^-. At least one such segment exists, and total their number is finite. By the symbol θ_α, we denote the segment which is first under the motion along D from $x_1 = -\infty$ to $x_1 = +\infty$. By the symbol $\theta(\alpha)$, we denote the length of this segment.

Each sufficiently general problem can be transformed into the such standard statement of a question. For example, under the research of a map $t = t(\zeta)$ of a domain G onto $|t| < 1$ a close finite point $\zeta = \zeta^*$, we transform the problem to the standard task with the substitution of variables

$$x_1 = \ln \frac{\zeta - \zeta'}{\zeta - \zeta^*}, \quad x_2 = \ln \frac{t - t(\zeta')}{t - t(\zeta^*)}$$

where ζ' and ζ^* are boundary points of G. This substitution transforms G onto a strip region D and a ball onto the strip.

The following result belongs to Ahlfors [2].

Theorem 7.1 *If $x' \in \theta_a$, $x'' \in \theta_b$, and*

$$\int_a^b \frac{d\alpha}{\theta(\alpha)} > 2,$$

then

$$y_1(x'') - y_1(x') > \pi \int_a^b \frac{d\alpha}{\theta(\alpha)} - 4\pi.$$

The following result belongs to Warschawski and permits to estimate $y_1(x'') - y_1(x')$ on the other side (see [195] – [197]). But in this case, a more severe constrains are required for C^+, C^-. Namely, we will assume that arcs C^+, C^- are described by

$$x_2 = \varphi^+(x_1), \quad x_2 = \varphi^-(x_1).$$

It is clear that

$$\theta(\alpha) = \varphi^+(\alpha) - \varphi^-(\alpha).$$

Besides, it is convenient to denote

$$\varphi(\alpha) = \frac{1}{2}\left[\varphi^+(\alpha) + \varphi^-(\alpha)\right],$$

and in this designation

$$\varphi^\pm(\alpha) = \varphi(\alpha) \pm \frac{1}{2}\theta(\alpha).$$

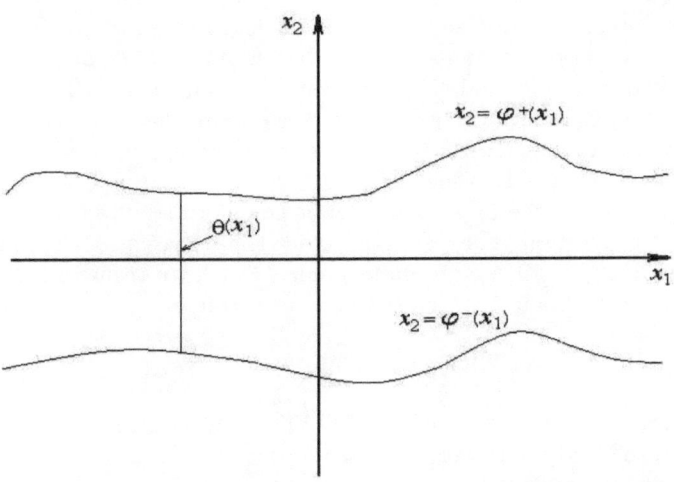

Fig. 7.1.

Theorem 7.2 *Suppose that for a constant $M > 0$ and every $-\infty < \alpha < +\infty$, it is fulfilled*

$$|\varphi'(\alpha)| < M, \quad |\theta'(\alpha)| < M.$$

If $a < b$, $x' \in \theta_a$, $x'' \in \theta_b$, then

$$y_1(x'') - y_1(x') <$$

$$< \pi \int_a^b \frac{1 + \varphi'^2(\alpha)}{\theta(\alpha)}\, d\alpha + \frac{\pi}{12} \int_a^b \frac{\theta'^2(\alpha)}{\theta(\alpha)}\, d\alpha + 12\pi(1 + M^2).$$

Our nearest aim is the obtaining of such results for conformal maps of a surface onto the plane strip.

7.2 Abutting Subdomains

Let G be a doubly-connected subdomain of $\mathbb{R}^2$. Let G_1 be a bounded and G_2 be an unbounded component of the set $\overline{\mathbf{R}}^2 \setminus G$. Denote by $\widetilde{\Gamma}$ the family of all Jordan curves lying in G and separating G_1 and G_2. The module $\text{mod}\,\widetilde{\Gamma}$ is called the *module of the ring domain G*. We denote it by the symbol $\mathrm{mod}\,(G)$. In this case, if G is a ring

$$\{(x_1, x_2) \in \mathbb{R}^2 : 0 < r < \sqrt{x_1^2 + x_2^2} < R < \infty\},$$

then by (1.32) we have

$$\mathrm{mod}\,(G) = \frac{1}{2\pi} \log \frac{R}{r}. \tag{7.3}$$

Now let D be a simply connected subdomain of $\mathbb{R}^2$ containing the point $(0,0)$ and having the nonempty boundary. If a number $r > 0$ is sufficiently small, then the set $D_r = \{x \in D : |x| > r\}$ is the doubly-connected domain. The function

$$\mathrm{mod}\,(D_r) + (2\pi)^{-1} \log r$$

is monotone increasing for $r \to 0$, and there exists a finite limit

$$\widetilde{\mathrm{mod}}\,(D) = \lim_{r \to 0} \left(\mathrm{mod}\,(D_r) + (2\pi)^{-1} \log r\right) \tag{7.4}$$

which is called a *reduced module of the domain D* with respect to the point $(0,0)$. If R_D be a inner conformal radius of a domain D with respect to $(0,0)$, then

$$\widetilde{\mathrm{mod}}\,(D) = (2\pi)^{-1} \log R_D \tag{7.5}$$

(see, for example, [66, Theorem **2.8**], [128]–[129], [93], [92]).

Let $D \subset \mathbb{R}^2$ be a simply connected domain with a nonempty boundary and let $\Omega \subset \mathbb{R}^m$ be a locally bi-Lipschitz surface of the form (2.1). Suppose that every prime end $e \in \partial\widetilde{D}$ has a principal point y_0 in which the surface Ω satisfies (4.27). By Theorem 4.24 it means that over every prime end $e \in \partial\widetilde{D}$, a unique prime end of Ω is arranged.

In the natural way, concepts of a simple Jordan arc (open or closed) and a simple Jordan curve are introduced in $\widetilde{D}$ and in $\widetilde{\Omega}$. In particular, the set of the prime ends $\partial\widetilde{D}$ is a simple Jordan curve in $\widetilde{D}$.

Let $E_1, E_2 \subset \widetilde{\Omega}$ be sets and $\gamma \subset \widetilde{\Omega}$ be a simple closed Jordan curve in $\widetilde{\Omega}$. We say that γ *separates* sets E_1, E_2 in $\widetilde{\Omega}$ if for every connected subset $K \subset \Omega$ closed in $\widetilde{\Omega}$ and such that $K \cap E_i \neq \emptyset$ $(i = 1, 2)$, we have $K \cap \gamma \neq \emptyset$.

Let $\Omega' \subset \Omega$ be a subdomain of the surface Ω, and let $e' \subset \partial\widetilde{\Omega}$ be a prime end. We say that the subdomain Ω' *abuts* to a prime end e' if for every sequence $\{\xi_n\}$ of points in Ω, convergent to e', there exists a number N such that for all $n > N$, the points ξ_n belong to the subdomain Ω'.

Let $\Omega \subset \mathbb{R}^m$ be a locally bi-Lipschitz surface given over a simply connected domain $D \subset \mathbb{R}^2$ with a nonempty boundary, $(0,0) \in D$, by a vector function (2.1). Let A and B be simply connected subdomains of D contained $(0,0)$. Let

K_1, K_2, ... be a sequence of continuums contained in A and in B and containing $(0,0)$. Suppose that all continuums K_n are such that $A_n = A \backslash K_n$, $B_n = B \backslash K_n$ are doubly-connected domains.

Lemma 7.6   *If diameters of K_n tend to 0, then there exists the following limit*

$$\lim_{n \to \infty} \left(\mathrm{mod}_\Omega(A_n) - \mathrm{mod}_\Omega(B_n) \right). \qquad (7.7)$$

This limit is not depend from the sequence of continuums $\{K_n\}$ with $\mathrm{diam}\, K_n \to 0$. In particular, if $\Omega = \mathbb{R}^2$, then

$$\lim_{n \to \infty} \left(\mathrm{mod}\,(A_n) - \mathrm{mod}\,(B_n) \right) = (2\pi)^{-1} \log(R_A/R_B) \qquad (7.8)$$

where R_A, R_B are inner conformal radii of domains A, B with respect to $(0,0)$.

Proof. We start from the statement (7.8). Let $\tilde{x} = F_n(x)$ be a schlicht conformal map of doubly-connected domains $A_n = A \setminus K_n$ onto doubly-connected domains $\widetilde{A}_n = \{ \tilde{x} \in A : |\tilde{x}| > \rho_n \}$ where numbers $\rho_n > 0$ are defined from relations

$$\mathrm{mod}\,(A_n) = \mathrm{mod}\,(\widetilde{A}_n), \quad n = 1, 2, \ldots$$

Moreover, we will assume that all $F_n(x)$ leave a fixed prime end e, of the domain A. The mappings $F_n(x)$ are defined by the unique way. The existence of such mappings follows from [47, Chapter V, §1, Theorem **2**].

Since diameters of continuums K_n tend to 0, then modules of doubly-connected domains $\mathrm{mod}\,(A_n)$ and $\mathrm{mod}\,(\widetilde{A}_n)$ tend to ∞, and numbers ρ_n to 0. From this we conclude that the sequence $F_1(x), F_2(x)$, ... tends to the identical map of A onto itself and, moreover, this convergence is uniform in every compact subset of the domain $A \setminus \{0\}$.

Indeed, let $w = \varphi(\tilde{x})$ be a schlicht conformal map of A onto $|w| < 1$ such that $\varphi(0) = 0$, $\varphi(e) = 1$. We put $\varphi_n = \varphi \circ F_n \circ \varphi^{-1}$ and denote by $\psi_n(w)$ the map, obtained by the symmetric extension of $\varphi_n(w)$ with respect to the unit circle. Because the schlicht maps $\psi_n(w)$ leave the point $w = 1$ fixed and do not take the values 0, ∞, then the sequence $\psi_1(w)$, $\psi_2(w)$, ... is a normal family [47, Chapter II, §7]. Let $\psi_{k_1}(w)$, $\psi_{k_2}(w)$, ... be its subsequence, uniformly convergent to a map $\varphi_0(w)$ uniformly inside $\mathbb{R}^2 \setminus \{0\}$. Since $|\psi_n(e^{i\theta})| = 1$, $0 \le \theta < 2\pi$, and the convergence is uniform on the circle $|w| = 1$, then $\psi_0(w) \not\equiv \mathrm{const}$. Therefore, the map $\psi_0(w)$ is schlicht in $\mathbb{R}^2 \setminus \{0\}$, and by the normalization

$$\psi_0(1) = 1, \quad \psi_0(0) = 0, \quad \psi_0(\infty) = \infty,$$

it coincides with the identical map. From here we conclude that maps $\varphi_n(w)$ converge uniformly to the identical map inside the ring $0 < |w| < 1$ and maps $F_n(x)$ inside the domain $A \setminus \{0\}$.

Fix a disk $B(0,d) = \{x \in \mathbb{R}^2 : |x| < d\}$ contained strongly inside every domain A and B. For sufficiently big n, the sets $D_n = D \setminus K_n$ are doubly-connected domains, and for the proof of (7.7), it is sufficient to establish the existence of the limit

$$\lim_{n\to\infty} (\mathrm{mod}\,(A_n) - \mathrm{mod}\,(D_n)) = (2\pi)^{-1} \log(R_A/R_D).\qquad(7.9)$$

Denote by p_n the image of the circle $|x| = d$ under the map $\widetilde{x} = F_n(x)$, and by $\widetilde{D}_n$ - the doubly-connected domain lying between p_n and the circle $|\widetilde{x}| = \rho_n$. Remark that $\widetilde{D}_n = F_n(D_n)$. The uniform convergence of $F_n(x)$ to the identical map implies that curves p_n converge uniformly to the circle $|\widetilde{x}| = d$. Therefore, we can assert that there exists a sequence $\varepsilon_1, \varepsilon_2, \ldots$ of positive numbers which converges to 0 such that every domain $\widetilde{D}_n$ is contained in the ring $\rho_n < |\widetilde{x}| < d + \varepsilon_n$ and also contains the ring $\rho_n < |\widetilde{x}| < d - \varepsilon_n$ inside itself. We compare modules of these rings with the module of the doubly-connected domain $\widetilde{D}_n$. Using (7.3) for the module of the ring, we arrive to the inequality

$$(2\pi)^{-1} \log \frac{d - \varepsilon_n}{\rho_n} \leq \mathrm{mod}\,(\widetilde{D}_n) \leq (2\pi)^{-1} \log \frac{d + \varepsilon_n}{\rho_n}\,.$$

From here we obtain

$$\lim_{n\to\infty} \left(\mathrm{mod}\,(\widetilde{D}_n) + (2\pi)^{-1} \log \rho_n\right) = (2\pi)^{-1} \log d\,.\qquad(7.10)$$

On the other hand, by (7.4) and (7.5) we have

$$\lim_{n\to\infty} \left(\mathrm{mod}\,(\widetilde{A}_n) + (2\pi)^{-1} \log \rho_n\right) = (2\pi)^{-1} \log R_A\,.\qquad(7.11)$$

Combining (7.10), (7.11) and taking into account that

$$d = R_D,\quad \mathrm{mod}\,(A_n) = \mathrm{mod}\,(\widetilde{A}_n),\quad \mathrm{mod}\,(D_n) = \mathrm{mod}\,(\widetilde{D}_n),$$

we arrive to (7.8).

The proof of the first statement of the lemma follows from Theorem 3.48. By Ω we find characteristics $(p(x), \theta(x))$ on the domain D. By the auxiliary quasi-conformal map $\xi = w(x) : D \to \mathbb{R}^2$, we introduce isothermal coordinates on Ω. Thus we define the simply connected domain $\mathcal{D} = w(D)$ and the metric

$$\lambda(\xi)\,|d\xi|^2,\quad \xi \in \mathcal{D},\qquad(7.12)$$

such that for every family of arcs (or curves) Γ lying in D, it is fulfilled

$$\mathrm{mod}_\Omega\,\Gamma = \mathrm{mod}_\lambda\,\Gamma^*$$

where $\Gamma^* = w(\Gamma)$ and $\mathrm{mod}_\lambda \Gamma^*$ means the module of Γ^* with respect to the metric (7.12). But the metric (7.12) is conformal, and hence,

$$\mathrm{mod}_\lambda \Gamma^* = \mathrm{mod}\,\Gamma^* . \qquad (7.13)$$

Thus we have

$$\mathrm{mod}_\Omega\,(A_n) - \mathrm{mod}_\Omega\,(B_n) = \mathrm{mod}\,(w(A_n)) - \mathrm{mod}\,(w(A_n)) . \qquad (7.14)$$

The existence of the limit (7.7) is equivalent to the existence of the limit in the left side of (7.14) and reduces to the proved statement. $\qquad\square$

We fix arbitrarily three prime ends $e', e_0, e'' \in \widetilde{D} \setminus D$, lying in the order of the positive circuit of the boundary $\widetilde{D} \setminus D$. Let $l \subset D$ be a Jordan arc separating the prime end e' from e_0 and e'' on D. Choose a locally Lipschitz function $h : D \to (0,1)$ with the following properties

$$\lim_{(x,y) \to e'} h(x,y) = 0, \quad h|_l = 1, \qquad (7.15)$$

and such that for every compact $A \subset \{(x,y) \in D : 0 < h(x,y) < 1\}$, it is fulfilled

$$\mathrm{ess}\inf_A |\nabla h(x,y)| > 0 . \qquad (7.16)$$

Denote by $E_h(t)$ the component of the set

$$\{(x,y) \in D : h(x,y) = t\}$$

separating the prime end e' from e_0 and e''. We put

$$\lambda_h(t) = \int_{E_h(t)} (g^{11}h_x^2 + 2g^{12}h_x h_y + g^{22}h_y^2)\, \frac{\sqrt{g}}{|\nabla h|}\, \sqrt{dx^2 + dy^2}$$

where g^{ij} $(i,j = 1,2)$ are elements of the inverse matrix $(g^{ij}) = (g_{ij})^{-1}$.

Lemma 7.17 *Let $f : D \to \mathbb{R}^2$ be a schlicht map conformal with respect to a metric ds_Ω^2. If*

$$\int_0 \frac{dt}{\lambda_h(t)} = \infty, \qquad (7.18)$$

then the image of the prime end e' is a prime end of the domain $f(D)$.

Proof follows from Length and Area Principle in the metric ds_Ω^2. Indeed, using (7.16) by (1.52), we can write

$$\int\limits_0^1 \mathrm{osc}^2(f, E_h(t)) \frac{dt}{\lambda_h(t)} \le$$

$$\le \iint_D \left(g^{11}|f_x|^2 + 2g^{12}\langle f_x, f_y \rangle + g^{22}|f_y|^2 \right) d\sigma_\Omega .$$

(7.19)

If $f(D) = \mathbb{R}^2$, then the boundary of $f(D)$ has the unique prime end, and then the statement is trivial.

Let $f(D) \ne \mathbb{R}^2$. Since the prime ends are invariant under conformal maps, then without loss of generality, we are right to assume that the domain $f(D)$ is the unit disc $B = B(0, 1)$. Under this assumption, we have

$$\iint_D \left(g^{11}|f_x|^2 + 2g^{12}\langle f_x, f_y \rangle + g^{22}|f_y|^2 \right) d\sigma_\Omega = \iint_D \sqrt{g}\, dx dy =$$

$$= 2\,\mathrm{area}\, f(D) = 2\pi.$$

To simplify calculations, it is sufficient to remark that the first one of double integrals is the Dirichlét integral for the map f which is conformal with respect the metric ds_Ω^2.

By (7.19) we have

$$\int\limits_0^1 \mathrm{osc}^2(f, E_h(t)) \frac{dt}{\lambda_h(t)} < \infty.$$

Now from (4.18) it follows that along a sequence of arcs $\{E_h(t_k)\}$, $k \to \infty$, it is fulfilled

$$\lim_{k \to \infty} \mathrm{osc}(f, E_h(t)) = 0.$$

Every arc $E_h(t_k)$ separates the prime end e' from the arc $\widetilde{e_0 e''}$. However,

$$\lim_{k \to \infty} \mathrm{diam}\, f(E_h(t_k)) = 0,$$

and for an arbitrary sequence of points $a_m \in D$, $a_m \to e'$, we can conclude that $\{f(a_m)\}$ converges to a point in ∂B. $\qquad\square$

Remark 7.20   It is easy to see that the property (7.18) characterizes the behavior of the metric ds_Ω^2 close to the prime end e' and does not depend from the arc l, separating e' from e_0 and e''.

$$\square$$

Below until the end of this chapter, we assume that the surface $\Omega \subset \mathbb{R}^m$ is locally bi-Lipschitz and is given by a vector function (2.1) over a simply connected domain $D \subset \mathbb{R}^2$ with the nonempty boundary. Let $G \subset D$ be a simply connected domain, $(0,0) \in G$. Suppose that Ω satisfies (7.18) at every prime end $e \in \partial G$ and that the function $\Lambda(x)/\lambda(x)$ is summable over the domain G.

Fix three different prime ends e', e_0, e'' of the domain G lying in the order of the positive circuit of the boundary $\widetilde{G} \setminus g$. Consider a simply connected subdomain G' abutting to the prime end e', not abutting to e'', and having the connected boundary $\partial_G G'$ with respect to the domain G.

Let γ_1, γ_2, ... be a chain of cross-sections of G defining the prime end e'. Suppose that it is so the chain of cross-sections of the subdomain G'. Let Γ'_n be a set of all locally rectifiable arcs $\gamma \subset G$, closings of which $[\gamma]_{\widetilde{G}}$ are simple Jordan arcs in $\widetilde{G}$, separating in $\widetilde{G}$ the cross-section $[\gamma_n]_{\widetilde{G}}$ and the closed arc $\widetilde{e_0 e''} \subset \widetilde{G} \setminus G$ which does not contain the prime end e'. Let Δ'_n be a set of all locally rectifiable arcs $\gamma \subset G'$ closings of which $[\gamma]_{\widetilde{G}}$ are simple Jordan arcs in $\widetilde{G}$ and separate in $\widetilde{G}$ sets $[\gamma_n]_{\widetilde{G}}$ and $[\partial_G G']_{\widetilde{G}}$. Since the subdomain G' abuts to e', then for sufficiently big n the sets Γ'_n and Δ'_n are nonempty.

Lemma 7.21   *There exists a finite limit*

$$\lim_{n\to\infty} \left(\mathrm{mod}_\Omega\left(\Gamma'_n\right) - \mathrm{mod}_\Omega\left(\Delta'_n\right)\right) \qquad (7.22)$$

which does not depend from the chain of cross-sections γ_1, γ_2, ... defining the prime end e'.

Proof. By (7.13) and the assumption on the summability of $\Lambda(x)/\lambda(x)$ over G, the map (2.1) is continuously extendable up to a homeomorphic map of $\tilde{\partial} D$ onto $\tilde{\partial}\Omega$. Thus, as in the proof of Lemma 7.6, it is sufficient to check (7.22) only in the case of the Euclidean metric $ds_\Omega = |dx|$.

Let $w = f(x)$ be a schlicht conformal map of the domain G onto the upper half-plane H_w^+ with

$$f(e') = 0, \quad f(e_0) = \alpha, \quad f(e'') = \infty \quad (\alpha > 0). \qquad (7.23)$$

Because a curves family module is a conformal invariant, then

$$\mathrm{mod}\left(\Gamma'_n\right) = \mathrm{mod}\left(f(\Gamma'_n)\right),$$

and by the symmetry principle for a curves family module in the Euclidean metric (see Lemma 1.20), we have

$$\text{mod}\,(\Gamma'_n) = 2\,\text{mod}\,\left(f(\Gamma'_n) + \overline{f(\Gamma'_n)}\right). \tag{7.24}$$

Close every arc of $\{f(\Gamma'_n) + \overline{f(\Gamma'_n)}\}$ adjoining its limiting points on the horizontal axis $\text{Im}\,w = 0$. By the symbol $[\Gamma'_n]$, we will denote the set of all arcs obtained by such procedure from $\{f(\Gamma'_n) + \overline{f(\Gamma'_n)}\}$.

It is easy to see that the curves family module does not change and the equality (7.24) is valid

$$\text{mod}\,(\Gamma_n) = 2\,\text{mod}\,[\Gamma'_n]. \tag{7.25}$$

Fix a cross-section γ_n of the cross-section chain defined the prime end e', and denote by $[\gamma_n]$ the arc obtained from the arc $f(\gamma_n) + \overline{f(\gamma_n)}$ with its closing. Let A_n be the doubly-connected domain lying between the arc $[\gamma_n]$ and the ray

$$L = \{w \in C_w : \text{Im}\,w = 0, \quad \text{Re}\,w \geq \alpha\}.$$

We show that

$$\text{mod}\,([\Gamma'_n]) = \text{mod}\,(A_n). \tag{7.26}$$

Indeed, every curve $\gamma \in [\Gamma'_n]$ lies in the domain A_n and separates boundary components A_n. Thus by the property of the module monotonicity, we have

$$\text{mod}\,([\Gamma'_n]) \leq \text{mod}\,(A_n). \tag{7.27}$$

On the other hand, since the family $f(\Gamma'_n) + \overline{f(\Gamma'_n)}$ consists of curves symmetric with respect to the real axis, then under calculations of its module, it is sufficient to restrict oneself to admissible functions $\rho(x)$ such that $\rho(x_1, x_2) = \rho(x_1, -x_2)$. Let $\rho(x)$ be a function admissible for $f(\Gamma'_n) + \overline{f(\Gamma'_n)}$ with such property and let γ be a curve separating boundary components of A_n. Denote by γ^+ the connected component of γ lying in the upper half-plane H_w^+ and separating in H_w^+ boundary components of A_n, by $\gamma^- \subset \gamma$ the arc with analogous property lying in the lower half-plane. Since A_n is symmetric with respect to the real axis, then curves $\gamma^+ + \overline{\gamma^+}$, $\gamma^- + \overline{\gamma^-}$ are contained in the family $f(\Gamma'_n) + \overline{f(\Gamma'_n)}$. From here by the symmetry of $\rho(x)$, we obtain

$$\int\limits_{\gamma} \rho(x)\,|dx| \geq \int\limits_{\gamma^+} + \int\limits_{\gamma^-} = \frac{1}{2}\int\limits_{\gamma^+ + \overline{\gamma^+}} + \frac{1}{2}\int\limits_{\gamma^- + \overline{\gamma^-}} \geq 1\,.$$

So, the function $\rho(x)$ is admissible for the family separating boundary components of A_n. Therefore,

$$\text{mod}\,\left(f(\Gamma'_n) + \overline{f(\Gamma'_n)}\right) \geq \text{mod}\,(A_n),$$

and
$$\mod\left(\left[\Gamma_n'\right]\right) \geq \mod\left(A_n\right). \tag{7.28}$$

Combining (7.27), (7.28), we arrive to (7.25), whence, taking into consideration (7.26), we obtain finally

$$\mod\left(\left[\Gamma_n'\right]\right) = 2\mod\left(A_n\right). \tag{7.29}$$

Denote by P_n the maximal of the doubly-connected domains in C_w, symmetric with respect to the real axis, coinciding with $f(G_n')$ in H_w^+ and having the curve $[\gamma_n]$ as a boundary component. Because the subdomain G' does not abut to the prime end e'' and has the connected boundary $\partial_G G'$, then the set of curves $[\Delta_n]$, obtained by closing from the family $f(\Delta_n') + \overline{f(\Delta_n')}$, consists of curves which lie in the domain P_n, symmetric with respect to $\operatorname{Im} w = 0$ and separate boundary components of P_n. As above, we prove that

$$\mod\left(\Delta_n'\right) = 2\mod\left(P_n\right). \tag{7.30}$$

The equalities (7.29), (7.30) and (7.8) imply the existence of the limit

$$\lim_{n\to\infty}\left(\mod\left(\Gamma_n'\right) - \mod\left(\Delta_n'\right)\right) = \pi^{-1}\log\frac{R_A}{R_P} \tag{7.31}$$

where R_A is the inner conformal radius with respect to the point $w = 0$ of the plane C_w with the cut along the way L, R_P is the inner conformal radius of the domain $P = \cup_{n=1}^{\infty}P_n \cup \{0\}$. The lemma is proved.   □

Analysis of relations (7.5), (7.8) and (7.31) suggests the existence of an analogy between the limit (7.22) and the reduced module of a simply connected domain with respect to a inner point. According to what had been said, we call the limit (7.22) the *reduced module* of the subdomain G' with respect to the prime end e' and the domain G with marked prime ends e_0, e''. Below we denote this limit by the symbol

$$k_\Omega(G', e'/e_0, e'')$$

or $k_\Omega(G')$ in the cases when it can not lead to misunderstandings.

We omit the subindex Ω in designations $k_\Omega(G', e'/e_0, e'')$ and $k_\Omega(G')$ for $\Omega = \mathbb{R}^2$.

The equality (7.31) contains essentially more information than it is necessary for the proof of this lemma. Since this surplus information is very important, we will describe the obtained result in detail.

Let $G \subset C_x$ be a simply connected domain with a nonempty boundary and fixed prime ends e', e_0, e''. Let G' be a simply connected subdomain G, abutting to e' and not abutting to e''. Let $w = f(x)$ be a schlicht conformal map of the domain G onto the upper half-plane H_w^+ with the normalization (7.23). Denote by P the maximal simply connected domain in C_w, containing the point $w = 0$, symmetric with respect to the real axis and coinciding with the domain $f(G')$ in H_w^+, by R_P its inner radius with respect to $w = 0$.

Lemma 7.32 *The following equality holds*

$$k(G', e'/e_0, e'') = \pi^{-1} \log \frac{4\alpha}{R_P} . \qquad (7.33)$$

The **proof** follows from (7.31), if we observe that $R_A = 4\alpha$. □

We single out the conformal invariance of the introduced quantity. Let $G \subset C_x$ be a simply connected domain with pointed prime ends e', e_0, e'' and let G' be its simply connected subdomain, abutting to e' and not abutting to e''. Let $\zeta = \varphi(x)$ be a schlicht conformal map of G onto a domain $U \subset C_\zeta$ with the boundary correspondence

$$\varphi(e') = t', \quad \varphi(e_0) = t_0, \quad \varphi(e'') = t''$$

where t', t_0, t'' are prime ends of U.

Lemma 7.34 *If U' is the image of G' under a map $\zeta = \varphi(x)$, then U' abuts to t' and does not abut to t'', and its reduced module $k(U', t'/t_0, t'')$ coincides with the reduced module $k(G', e'/e_0, e'')$.*

Proof. By the Carathéodory theory, the conformal map $\zeta = \varphi(x)$ induces a homeomorphism of the set of prime ends of G onto the set of prime ends of the domain U. Moreover, every chain of cross-sections of G transforms onto a chain of cross-sections of U and conversely. From here it follows that the subdomain U' abuts to the prime end t' and does not abut to t''.

The construction of reduced modules $k(G')$ and $k(U')$ is based on the conformally invariant quantities and does not depend from chains of cross-sections defined prime ends e' and t'. Thus the equality of $k(G')$ and $k(U')$ is obvious.
□

Now we suppose that G' and G'' are arbitrary simply connected subdomains of a simply connected domain $G \subset \Omega$ abutting to prime ends e' and e'' respectively. Suppose that $G' \cap G'' = \emptyset$ and that $\partial_G G'$ and $\partial_G G''$ are connected.

We fix a prime end e_0 of G. Then the reduced modules $k_\Omega(G', e'/e_0, e'')$ and $k_\Omega(G'', e'/e_0, e'')$ are defined.

Lemma 7.35   *The following inequality holds*

$$k_\Omega(G') + k_\Omega(G'') \geq \pi^{-1} \log 16 \, . \tag{7.36}$$

Proof. It is sufficient to check the relation (7.36) in the case $\Omega = \mathbb{R}^2$. By Lemma 7.34 reduced modules $k(G')$ and $k(G'')$ are conformally invariant, and it is sufficient to consider the situation in which the domain G coincides with the upper half-plane H_x^+, and prime ends e', e_0, e'' are identical to boundary points 0, α, ∞ ($\alpha > 0$) respectively. We use Lemma 7.32 setting $f(x) \equiv x$. We keep the previous sense for designations P and R_P and denote by Q the maximal simply connected domain in $\overline{C}_w$, containing the point $w = \infty$, symmetric with respect to the real axis $\operatorname{Im} w = 0$ and coinciding with $f(G'')$ in H_w^+. By Lemma 7.32

$$k(G') = \pi^{-1} \log \frac{4\alpha}{R_P} \, . \tag{7.37}$$

We show that

$$k(G'') = \pi^{-1} \log \frac{4}{\alpha R_Q} \tag{7.38}$$

where R_Q is the inner conformal radius of the domain Q with respect to the point $w = \infty$.

Let $\tilde{w} = F(w)$ be a schlicht conformal map of the half-plane H_w^+ onto itself with the boundary correspondence

$$F(\infty) = 0, \quad F(\alpha) = \alpha, \quad F(0) = \infty \, .$$

Let U be a maximal domain containing the point $\tilde{w} = 0$, symmetric with respect to the real axis and coinciding with $F(G'')$ in $H_{\tilde{w}}^+$. Let R_U be its inner conformal radius with respect to the point $\tilde{w} = 0$. By Lemma 7.32

$$k(G'') = \pi^{-1} \log \frac{4\alpha}{R_U} \, .$$

But $\tilde{w} = F(w)$ is the restriction of the map $\varphi(w) = \frac{4\alpha}{w}$ onto H_w^+. Observing that $U = \varphi(Q)$, $R_U = \alpha^2 R_Q$, we obtain the necessary statement.

Further, since subdomains G' and G'' are non-overlapping in H_x^+, then domains P and Q are non-overlapping also. Using the inequality

$$R_P \cdot R_Q \leq 1 \, , \tag{7.39}$$

connecting conformal radii of not leaning domains [66, Theorem **7.1**], and taking into account (7.37), (7.38), we verify that the lemma is valid.   □

We make some remarks about the possibility of the equality in (7.36). We restrict ourselves with the case of the Euclidean metric. At first, we remark that the equality in (7.39) is possible if and only if the domain P in C_w is the disc with the center at the origin, and the domain $Q = \overline{C}_w \setminus \overline{P}$ [66]. Suppose that the domain $G \subset C_x$ coincides with H_x^+ and prime ends e', e_0, e'' are identical to boundary points 0, α, ∞. From the analysis of the proof of the lemma, we see that the equality in (7.36) is possible if and only if the subdomain $G' \subset H_x^+$ is the half-disc

$$K' = \{x \in H_x^+ : |x| < r\}, \qquad G'' = H_x^+ \setminus \overline{K'}.$$

In the case of an arbitrary domain G, the equality in (7.36) is possible if there exists a schlicht conformal map of G onto H_w^+ with the normalization (7.23) under which G' is mapping onto K', and G'' onto $H_w^+ \setminus \overline{K'}$.

7.3   Estimates of Conformal Maps

The introduced quantity permits to reformulate some boundary problems of the theory of conformal mappings of surfaces to inner problems of the theory of conformal maps of plane domains for which methods of solutions are detailed today. As an example, we consider three following problems, connected with theorems of the theorem type 7.1 and 7.2 about conformal maps of a plane domain onto the plane.

Let $G \subset \Omega$ be a domain with fixed prime ends e', e_0, e'' and let $\zeta = F(x)$ be a schlicht conformal map of G onto the strip[1]

$$\Pi = \{\zeta \in C : 0 < \operatorname{Im}\zeta < \pi\}$$

with the normalization

$$\lim_{x \to e'} \operatorname{Re} F(x) = -\infty, \quad F(e_0) = 0, \quad \lim_{x \to e''} \operatorname{Re} F(x) = +\infty. \tag{7.40}$$

For every simply connected subdomain $G' \subset G$ abutting to the prime end e', we put

$$I_1(G'; F) = \inf_{x \in \sigma} \operatorname{Re} F(x) \quad (\sigma = G \setminus G').$$

---

[1]Below it is convenient to assume that $\zeta = F(x)$ is a complex valued function where the argument is the point $x \in \Omega$.

If G'' abuts to e'' and does not overlap on G', then let

$$I_2(G', G''; F) \;=\; \inf_{x \in \sigma} \operatorname{Re} F(x),$$

$$I_3(G', G''; F) \;=\; \sup_{x \in \sigma} \operatorname{Re} F(x), \quad (\sigma = G \setminus (G' \cup G'')),$$

$$I_4(G', G''; F) \;=\; \sup_{x', x'' \in \sigma} |\operatorname{Re} F(x') - \operatorname{Re} F(x'')|.$$

We consider the following problems.

Problem A. *Let* $k_\Omega(G', e'/e_0, e'')$ *be a given reduced module. It is necessary to find the infimum of* $I_1(G'; F)$.

Problem B. *Let* $k_\Omega(G', e'/e_0, e'')$ *and* $k_\Omega(G'', e''/e', e_0)$ *be given reduced modules. Find the infimum and the supremum of* $I_2(G', G''; F)$ *and* $I_3(G', G''; F)$ *respectively.*

Problem C. *Suppose that the assumptions of the problem B hold. It is necessary to find the supremum of* $I_4(G', G''; F)$.

Each one of these boundary problems is equivalent to an inner problem. Indeed, with an auxiliary map $w = e^\zeta$ we transform this situation into the research of the map $f(x) = e^{F(x)}$ of the domain $G \subset \Omega$ onto the half-plane H_w^+. The normalization (7.40) of $F(x)$ guarantees the normalization (7.23) with $\alpha = 1$ of the map $f(x)$.

Denote by P the maximal simply connected domain in the complex plane C_w, containing the point $w = 0$, symmetric with respect to the real axis and coinciding with $f(G')$ in the half-plane H_w^+; by Q we denote the maximal domain in the extended plane $\overline{C}_w$, containing $w = \infty$, symmetric with respect to the real axis and coinciding with $f(G'')$ in H_w^+. From (7.33) we find

$$R_P = 4 \exp\{-\pi\, k(G')\}; \tag{7.41}$$

and, reasoning as in the proof of the equality (7.38), we obtain

$$R_Q = 4 \exp\{-\pi\, k(G'')\}. \tag{7.42}$$

Thus problems **A**, **B** and **C** are equivalent to the following problems.

Problem A′. *Let* $P \subset C_w$ *be an arbitrary simply connected domain, symmetric with respect to the axis* $\operatorname{Im} w = 0$, *containing the point* $w = 0$, *and having a given inner conformal radius* R_P *with respect to* $w = 0$. *We need to find*

$$\inf_{w \in \Sigma} |w| \quad (\Sigma = C_w \setminus P).$$

Problem B′. *Let P and Q be an arbitrary pair of non-overlapping simply connected domains in $\overline{C}_w$, containing points $w = 0$ and $w = \infty$ respectively, symmetric with respect to the real axis, and having inner conformal radii R_P, R_Q with respect to $w = 0$ and $w = \infty$. We need to find*

$$\inf_{w \in \Sigma} |w|, \quad \sup_{w \in \Sigma} |w| \quad (\Sigma = C_w \setminus (P \cup Q)).$$

Problem C′. *Suppose that the assumptions of the problem B′ hold. We need to find*

$$\sup_{w', w'' \in \Sigma} \frac{|w'|}{|w''|} \quad (\Sigma = C_w \setminus (P \cup Q)).$$

The solution of the problem **A′** is given by the Koebe theorem about $\frac{1}{4}$ (see, for example, [47, Chapter II, §4]). Hence, by (7.41) we have the following statement.

Theorem 7.43 *Under assumptions of the problem* **A**, *the following inequality holds*

$$I_1(G'; F) > -\pi \, k_\Omega(G'). \tag{7.44}$$

If $k_\Omega(G')$ is fixed, then the inequality can not be improved, .

For the **proof** it is sufficient to check that the estimate (7.44) is unimprovable. Here it is necessary to remark that the extremal domain in the Koebe theorem is the plane cut along a ray and that there exist domains P with the given conformal radius R_P, symmetric with respect to Im $w = 0$, having Jordan boundaries and arbitrarily close to the ray domain of Koebe. $\qquad\square$

We will describe the solution of the problem **B′** given by Teichmüller [183].[2]

Lemma 7.45 *For an arbitrary point $w \in \Sigma$, we have*

$$\mu(t_0) \, R_P/4 \le |w| \le 4/(R_Q \mu(t_0)) \tag{7.46}$$

where t_0 is the unique root of the equation

$$\nu(t) = R_P \cdot R_Q/16 \tag{7.47}$$

and functions $\mu(t)$, $\nu(t)$ are defined by relations:

$$\mu(t) = [(t^2 - 1)/t] \cdot [(t+1)/(t-1)]^{1/t}, \tag{7.48}$$

$$\nu(t) = [t^2/(t^2 - 1)^2] \cdot [(t-1)/(t+1)]^{t+1/t}. \tag{7.49}$$

---

[2]This result can be extracted from the article of Kühnau [81].

Equalities in (7.46) are valid simultaneously if and only if $R_P \cdot R_Q = 1$ or, what is the same, if the set Σ is a circle with the center at $w = 0$.

If $R_P \cdot R_Q < 1$, then the full description of extremal domains where estimates (7.46) are realized, is very cumbersome. Therefore, we restrict ourselves to their qualitative description.

The left inequality in (7.46) is a equality if and only if the domain P belongs to the special family of domains $\mathcal{M}$, $Q = \overline{C}_w \setminus \overline{P}$. Every domain of the set $\mathcal{M}$ is symmetric with respect to the real axis and can be obtained from a domain with a Jordan boundary by the carrying out a cut of a positive length lying on the real axis. Moreover, the lower inequality in (7.46) is achieved at the end point of this cut.

The right inequality in (7.46) is a equality if and only if the domain Q belongs to the family of domains $\mathcal{N}$, $P = C_w \setminus \overline{Q}$; every domain of the set $\mathcal{N}$ can be obtained as the image of a domain of $\mathcal{M}$ under an inversion with respect to a circle with the center at the origin. The upper estimate in (7.46) is achieved also in the end point of the cut.

Lemma 7.45 gives the full solution of the problem **B**.

Theorem 7.50   *Under assumptions of the problem* **B** *the following equalities hold:*

$$I_2(G', G''; F) \geq -\pi\, k_\Omega(G') + \log \mu(t_0)\,, \tag{7.51}$$

$$I_3(G', G''; F) \leq \pi\, k_\Omega(G'') - \log \mu(t_0) \tag{7.52}$$

where t_0 is the unique root of the equation

$$\pi^{-1} \log \frac{1}{\nu(t)} = k_\Omega(G') + k_\Omega(G'') \tag{7.53}$$

and functions $\mu(t)$, $\nu(t)$ are defined by (7.48), (7.49).
   Equalities in (7.51), (7.52) are reached if and only if

$$k_\Omega(G') + k_\Omega(G'') = \pi^{-1} \log 16\,. \tag{7.54}$$

In other cases, estimates are strict, but for fixed $k_\Omega(G')$ and $k_\Omega(G'')$, they can not be replaced with better estimates.

Proof. Inequalities (7.51) and (7.52) follow from (7.46). We consider in detail possibilities of equalities in these estimates. At first we assume that the condition (7.54) holds. The relation (7.53) implies

$$\nu(t_0) = \frac{1}{16}\,, \quad t_0 = 1, \quad \mu(t_0) = 4\,;$$

and (7.51), (7.52) are transformed to the form

$$I_2(G', G''; F) = I_3(G', G''; F) = -\pi k_\Omega(G') + \log 4 = \pi k_\Omega(G'') - \log 4 \,.$$

Now we suppose that (7.54) is not true, but there is equality in (7.51). Then there exists a point $w_0 \in \Sigma$ in which the lower estimate in (7.46) is achieved. Thus the domain P belongs to the set $\mathcal{M}$, $Q = \overline{C}_w \setminus P$. However, P can be obtained from a domain with a Jordan boundary by a cut along the real axis where w_0 is the end point of the cut. The point w_0 should lie in the positive distance from the image of the set $G \setminus (G' \cup G'')$ under the map $w = f(x)$, what is impossible.

By analogous way, we prove the rigor of the estimate (7.52) if the condition (7.54) is not true.

Unimprovability of (7.51), (7.52) for fixed modules $k_\Omega(G')$, $k_\Omega(G'')$ follows from unimprovability of inequalities (7.46) in the problem $\mathbf{B}'$ for the subclass of pairs of domains with Jordan boundaries. $\qquad\square$

Corollary 7.55   *Under assumptions of the problem $\mathbf{B}$, it is fulfilled*

$$-\pi \, k_\Omega(G') < I_2(G', G''; F) \le I_3(G', G''; F) < \pi \, k_\Omega(G'') \,. \qquad (7.56)$$

The **proof** follows from (7.51), (7.52) and the inequality $\mu(t) > 1$. $\qquad\square$

We do not known the precise solution of the problem $\mathbf{C}'$. Therefore, for the estimate of $I_4(G', G''; F)$, we will use Theorem 7.50 and show that the obtained inequality is not too different from the best.

Theorem 7.57   *Under assumptions of the problem $\mathbf{C}$, it is fulfilled*

$$\pi \, \delta(G', G'') - \log 16 \le I_4(G', G''; F) \le \pi \, \delta(G', G'') - \log \mu^2(t_0) \qquad (7.58)$$

where t_0 is a root of (7.53), and

$$\delta(G', G'') = k_\Omega(G') + k_\Omega(G'') \,.$$

Proof. The upper estimate in (7.58) is the corollary of (7.51) and (7.52). We check the lower estimate. Passing to the plane C_w, reasoning as above and using by Theorem 1 [88, Chapter III, §1] about non-overlapping domains, we have

$$\frac{1}{R_Q} \le \sup_{w \in \Sigma} |w| \,, \quad \inf_{w \in \Sigma} |w| \le R_p \,.$$

From this, taking into account (7.41) and (7.42), we obtain

$$\pi \, k_\Omega(G'') - \log 4 \le \sup_{x \in \sigma} \operatorname{Re} F(x) \,, \quad \inf_{x \in \sigma} \operatorname{Re} F(x) \le -\pi \, k_\Omega(G') + \log 4$$

that implies to the lower estimate in (7.58).   □

If the assumption (7.54) holds, then inequalities (7.58) are precise equalities. Otherwise, evidently, the upper estimate in (7.58) is not precise. However, it is not too different from the best estimate, because $1 < \mu(t) \le 4$ for all t.

7.4   Estimates of $k_\Omega(G')$, $k_\Omega(G'')$

Below we provide estimates of the reduced modules

$$k_\Omega(G', e'/e_0, e''), \quad k_\Omega(G'', e''/e', e_0)$$

with modules of some special families of arcs. Moreover, we show connections of the introduced quantities with known characteristics.

Let $G \subset \Omega$ be a simply connected domain with the fixed prime ends e', e_0, e''. Let G' and G'' be simply connected subdomains of G, abutting to e', e'', respectively, and not overlapping. Fix a chain of cross-sections γ_1, γ_2, ... of G, defining the prime end e' and being a chain of cross-sections of G'. Let Γ' and Δ' be arc families, described in the definition of the reduced module $k_\Omega(G')$. Moreover, let Λ' be a family of locally rectifiable arcs $\gamma \subset G \setminus \overline{G'}$ closings of which in $[\gamma]_{\widetilde{G}}$ are simple Jordan arcs in $\widetilde{G}$, separating $[\partial_G G']_{\widetilde{G}}$ and $\widetilde{e_0 e''}$, and $\mathcal{E}'_n$ be a family of arcs $\gamma \subset G$ closings of which $[\gamma]_{\widetilde{G}}$ separate $[\gamma_n]_{\widetilde{G}}$, $\widetilde{e_0 e''}$ and such that $[\gamma]_{\widetilde{G}} \cap \widetilde{e_0 e''} \ne \emptyset$. Observe that if $[\partial_G G']_{\widetilde{G}} \cap \widetilde{e_0 e''} \ne \emptyset$, then the set $\Lambda' = \emptyset$.

Lemma 7.59   *If the boundary $\partial_G G'$ is connected, then*

$$\Gamma'_n = \Delta'_n \cup \Lambda' \cup \mathcal{E}'_n. \tag{7.60}$$

Proof. Let $\gamma \in \Gamma'_n$ be an arc. If $[\gamma]_{\widetilde{G}} \cap [\partial_G G']_{\widetilde{G}} \ne \emptyset$, then $\gamma \in \mathcal{E}'_n$. Therefore, we assume that $[\gamma]_{\widetilde{G}} \cap [\partial_G G']_{\widetilde{G}} = \emptyset$. Then, γ is containing in G' or in $G \setminus \overline{G'}$. We consider every case separately.

Let $\gamma \subset G'$. We show that $\gamma \in \Delta'_n$. Setting the opposite, i.e. that the arc $[\gamma]_{\widetilde{G}}$ does not separate γ_n and $[\partial_G G']_{\widetilde{G}}$, we find a continuum $K \subset \widetilde{G}$, joining $[\gamma_n]_{\widetilde{G}}$ with $[\partial_G G']_{\widetilde{G}}$ and such that $[\gamma]_{\widetilde{G}} \cap K = \emptyset$. Since the boundary $\partial_G G'$ is connected, then the set $G \setminus G'$ is also connected, and from $G'' \subset (G \setminus G')$, it follows that $[G \setminus G']_{\widetilde{G}} \cap \widetilde{e_0 e''} = \emptyset$. Thus, the set $K \cup [G \setminus G']_{\widetilde{G}}$ is connected and joints the arc $[\gamma_n]_{\widetilde{G}}$ with the arc $\widetilde{e_0 e''}$. Thus, we arrive to the contradiction with $\gamma \in \Gamma'_n$.

So, every arc $\gamma \in \Gamma'_n$ belongs to one of the families $\mathcal{E}'_n$, Δ'_n, Λ', and we proved the enclosure

$$\Gamma'_n \subset (\Delta'_n \cup \Lambda' \cup \mathcal{E}'_n).$$

For the proof of the opposite enclosure, it is sufficient to prove that $\Lambda' \subset \Gamma'_n$. Let $\Lambda' \neq \emptyset$ and $\gamma \in \Lambda'$ be an arc. Assuming that $\gamma \notin \Gamma'_n$, we find a continuum $K \subset \widetilde{G}$, joining $[\gamma_n]_{\widetilde{G}}$, $\widetilde{e_0e''}$ in $\widetilde{G}$ and not intersecting $[\gamma]_{\widetilde{G}}$. If $K \cap [\partial_G G']_{\widetilde{G}} \neq \emptyset$, then we have the contradiction to the hypothesis $\gamma \in \Lambda'$. Thus, let $K \cap [\partial_G G']_{\widetilde{G}} = \emptyset$. There are two possibilities: either $[\partial_G G']_{\widetilde{G}} \cap \widetilde{e_0e''} \neq \emptyset$, or this intersection is empty. In the first case, we arrive at the contradiction to the assumption that $\Lambda' \neq \emptyset$. In the second case, we consider the set $K \cup \widetilde{e_0e''}$. This set is connected and joins cross-sections $[\gamma_n]_{\widetilde{G}}$ with the prime end e'' but does not intersect $[\partial_G G']_{\widetilde{G}}$. It is not possible, because subdomains G' and G'' are non-overlapping. The lemma is proved. $\qquad\square$

Keeping conditions and designations of the previous section, we will denote by $\mathcal{E}'$ the family of all arcs $\gamma \subset G'$, closings of which are simple Jordan arcs in $\widetilde{G}$ separating in $\widetilde{G}$ the prime end e' and the arc $\widetilde{e_0e''}$ such that $[\gamma]_{\widetilde{G}} \cap [\partial_G G']_{\widetilde{G}} \neq \emptyset$.

Lemma 7.61 *The following inequality holds*

$$\operatorname{mod}_\Omega (\Lambda') \leq k_\Omega(G', e'/e_0, e'') \leq \operatorname{mod}_\Omega (\Lambda') + \operatorname{mod}_\Omega (\mathcal{E}'). \qquad (7.62)$$

Proof. Since families Λ' and Δ'_n lie in nonintersecting measurable sets, by (1.18) we have

$$\operatorname{mod}_\Omega (\Delta'_n \cup \Lambda') = \operatorname{mod}_\Omega (\Delta'_n) + \operatorname{mod}_\Omega (\Lambda').$$

But by (7.60) it is fulfilled $(\Delta'_n \cup \Lambda') \subset \Gamma'_n$, and using the property of the module monotonicity

$$\operatorname{mod}_\Omega (\Delta'_n \cup \Lambda') \geq \operatorname{mod}_\Omega (\Gamma'_n),$$

we obtain

$$\operatorname{mod}_\Omega (\Lambda') \leq \operatorname{mod}_\Omega (\Gamma'_n) - \operatorname{mod}_\Omega (\Delta'_n).$$

Passing to the limit, we verify the lower estimate in (7.62).

We prove the upper estimate. From (7.60) and (1.18), it follows that

$$\operatorname{mod}_\Omega (\Gamma'_n) \leq \operatorname{mod}_\Omega (\Delta'_n) + \operatorname{mod}_\Omega (\Lambda') + \operatorname{mod}_\Omega (\mathcal{E}'_n).$$

However, $\mathcal{E}'_n \subset \mathcal{E}'$ and

$$\operatorname{mod}_\Omega (\mathcal{E}'_n) \leq \operatorname{mod}_\Omega (\mathcal{E}').$$

Therefore,

$$\operatorname{mod}_\Omega (\Gamma'_n) - \operatorname{mod}_\Omega (\Delta'_n) \leq \operatorname{mod}_\Omega (\Lambda') + \operatorname{mod}_\Omega (\mathcal{E}'),$$

and, passing to the limit, we obtain the necessary statement. $\qquad\square$

Combining (7.44) and (7.62), we arrive to the theorem.

Theorem 7.63   *Let $\zeta = F(x)$ be a schlicht conformal map of a domain $G \subset \Omega$ onto a strip Π normalized with conditions (7.40), and let G' be a subdomain of G abutting to the prime end e'.*

Then for all $x \in G \setminus G'$, the following inequality holds

$$\operatorname{Re} F(x) > -\pi \left\{ \operatorname{mod}_\Omega (\Lambda') + \operatorname{mod}_\Omega (\mathcal{E}') \right\}. \qquad (7.64)$$

Keeping designations Λ', $\mathcal{E}'$, we denote by Λ'', $\mathcal{E}''$ corresponding families of arcs which appear under the consideration of the subdomain G''. Moreover, let Λ be the family of arcs $\gamma \subset G$ closings of which $[\gamma]_{\widetilde{G}}$ are simple Jordan arcs in $\widetilde{G}$ separating sets $[G']_{\widetilde{G}}$ and $[G'']_{\widetilde{G}}$ in $\widetilde{G}$. For an arbitrary set $K \subset \widetilde{G}$, let $\mathcal{E}(K)$ be the family of arcs $\gamma \subset G$ separating prime ends e', e'' and such that $[\gamma]_{\widetilde{G}} \cap [K]_{\widetilde{G}} \neq \emptyset$.

Lemma 7.65   *The following estimate holds*

$$\operatorname{mod}_\Omega (\Lambda) \leq k_\Omega(G') + k_\Omega(G'') \leq \operatorname{mod}_\Omega (\Lambda) + S \qquad (7.66)$$

where

$$S = \operatorname{mod}_\Omega(\mathcal{E}_1) + \operatorname{mod}_\Omega(\mathcal{E}_2) + \min\{\operatorname{mod}_\Omega(\mathcal{E}_1), \operatorname{mod}_\Omega(\mathcal{E}_2)\}$$

and

$$\mathcal{E}_1 = \mathcal{E}(\partial_G G'), \quad \mathcal{E}_2 = \mathcal{E}(\partial_G G'').$$

Proof. It is not difficult to see that (7.37), (7.38) implies that the quantity $k_\Omega(G') + k_\Omega(G'')$ does not depend on the prime end e_0. We choose e_0 as a prime end of $[\partial_G G'']_{\widetilde{G}}$ lying between prime ends e', e'' in the order of the positive direction of the circuit of the boundary $\widetilde{G} \setminus G$. The family $\Lambda'' = \emptyset$ and by Lemma 7.61, we have

$$0 \leq k_\Omega(G'') \leq \operatorname{mod}_\Omega (\mathcal{E}'').$$

However,

$$\mathcal{E}'' \subset \mathcal{E}_2 = \mathcal{E}(\partial_G G''),$$

and

$$0 \leq k_\Omega(G'') \leq \operatorname{mod} (\mathcal{E}_2). \qquad (7.67)$$

Now we remark that

$$\Lambda \subset \Lambda' \subset (\Lambda \cup \mathcal{E}_2),$$

and hence,

$$\operatorname{mod}_\Omega(\Lambda) \leq \operatorname{mod}_\Omega(\Lambda') \leq \operatorname{mod}_\Omega(\Lambda) + \operatorname{mod}_\Omega(\mathcal{E}_2).$$

Using (7.62), (7.67) and taking into account that

$$\mathrm{mod}_\Omega(\mathcal{E}') \leq \mathrm{mod}_\Omega(\mathcal{E}_1)\,,$$

we have

$$\mathrm{mod}_\Omega(\Lambda) \leq k_\Omega(G') + k_\Omega(G'') \leq \qquad (7.68)$$

$$\leq \mathrm{mod}_\Omega(\Lambda) + \mathrm{mod}_\Omega(\mathcal{E}_1) + 2\,\mathrm{mod}_\Omega(\mathcal{E}_2)\,.$$

On the other hand, choosing e_0 as a prime end, belonging to the set $[\partial_G G']_{\widetilde{G}}$, and lying between e', e'', we easily arrive to the inequality

$$\mathrm{mod}_\Omega(\Lambda) \quad \leq k_\Omega(G') + k_\Omega(G'') \leq$$

$$\leq \mathrm{mod}_\Omega(\Lambda) + \mathrm{mod}_\Omega(\mathcal{E}_2) + \mathrm{mod}_\Omega(\mathcal{E}_1)\,. \qquad (7.69)$$

Comparing (7.68) and (7.69), we obtain the necessary statement. $\qquad\square$

The following lemma is a sharpening of the inequality (7.66) in the special case if $\partial_G G' = \partial_G G''$.

Lemma 7.70 *Suppose that a Jordan arc $L \subset G$ separates G into two subdomains G' and G'' such that the subdomain G' abuts to the prime end e' and the subdomain G'' to the prime end e''. Then*

$$0 \leq k_\Omega(G') + k_\Omega(G'') \leq 2\,\mathrm{mod}_\Omega\left(\mathcal{E}(L)\right)\,. \qquad (7.71)$$

As in the **proof** of Lemma 7.65, we fix a prime end $e_0 \in [L]_{\widetilde{G}}$. Then $\Lambda' = \Lambda'' = \emptyset$. Remarking that $\mathcal{E}', \mathcal{E}'' \subset \mathcal{E}(L)$ and using (7.62), we obtain the necessary statement. $\qquad\square$

Inequalities (7.62), (7.66), (7.71) imply from Theorems 7.43 – 7.57 series of estimates of conformal mappings of a surface onto the strip which are less precise, but written in the terms of customary quantities. We bring two of such statements.

Theorem 7.72 *Let $\zeta = F(x)$ be a schlicht conformal map of a domain $G \subset \Omega$ onto the strip Π normalized with conditions (7.40), and let $L \subset G$ be a Jordan arc separating prime ends e', e''.*
Then

$$-\pi \left\{\mathrm{mod}_\Omega(\Lambda') + \mathrm{mod}_\Omega(\mathcal{E}(L))\right\} \leq \mathrm{Re}\, F(x) \quad (\forall\, x \in L)\,, \qquad (7.73)$$

and

$$\mathrm{osc}_{x \in L}\, \mathrm{Re}\, F(x) \leq 2\pi\,\mathrm{mod}_\Omega(\mathcal{E}(L)) \qquad (7.74)$$

where Λ' is the family of curves, separating L and $\widetilde{e_0 e''}$, and

$$\mathrm{osc}_{x \in L} \operatorname{Re} F(x)$$

means the oscillation of $\operatorname{Re} F(x)$ along the set L.

Proof. The inequality (7.73) follows from (7.44) and (7.62), the inequality (7.74) is the corollary of (7.58) and (7.71). $\qquad\qquad\qquad\qquad\qquad\qquad\square$

Combining inequalities (7.51), (7.52), (7.62), we arrive to upper estimates for $\operatorname{Re} F(x)$ in the set L. For example, remark the following estimate.

Theorem 7.75 *Under conditions of Theorem 7.72, it is fulfilled*

$$\operatorname{Re} F(x) \leq \pi \{ \mathrm{mod}_\Omega(\Lambda'') + \mathrm{mod}_\Omega(\mathcal{E}'') \} - \log \mu(t_0) \quad (x \in L) . \tag{7.76}$$

Other estimates in the case of the Euclidean metric, see in [103], [18].

Open questions 7.77   1) It is interesting to find estimates of $k(G')$, $k(G'')$ with geometric quantities, different from the module of arc families.   2) It is desirable to find analogues of Theorems 7.1 and 7.2 for strip regions on surfaces with curvature restrictions.

Chapter 8

The Stabilization Speed of Solutions

We show bounds of the admissible stabilization speed of the gas dynamic equation solutions under excess of which, solutions can be only identical constants [124]. Proofs are based on estimates of Ahlfors-Warschawski type theorems considered in the previous chapter. About the general setting of the problem, see [114], [144], [5], [200], [201], [142], [169], [48] etc.

8.1   The Gas Dynamics Equation

Below we consider generalization solutions of equations in the form

$$\frac{\partial}{\partial x}\left(\delta(q)\varphi_x\right) + \frac{\partial}{\partial y}\left(\delta(q)\varphi_y\right) = 0, \quad q = |\nabla\varphi| \tag{8.1}$$

where $\delta(q)$ is a continuous function, $\delta(0) = 1$.

In the case

$$\delta(q) = \left(1 - \frac{\gamma - 1}{2}q^2\right)^{1/(\gamma-1)},$$

we have the classic gas dynamic equation. This equation describes the speed potential of a plane established stream of an ideal gas in the adiabatic duty γ, $-\infty < \gamma < +\infty$, is a constant, characterizing the gas (see, for example, [20, Chapter I, §**2**] or [86, Chapter IV, §**15**]). For $\gamma = 1 \pm 0$, we put

$$\delta(q) = \exp\left\{-\frac{1}{2}q^2\right\}.$$

This equation has the elliptic type for $\gamma \leq 1$. For $\gamma > 1$ it is elliptic if $q < \sqrt{2/(\gamma - 1)}$, parabolic if $q = \sqrt{2/(\gamma - 1)}$, and hyperbolic if $q > \sqrt{2/(\gamma - 1)}$. Below we assume that $\gamma \leq 1$, or $\gamma > 1$ and

$$\underset{D'}{\text{ess sup}}\, q(x, y) < \sqrt{\frac{2}{\gamma - 1}} \text{ for every subdomain } D' \subset\subset D, \qquad (8.2)$$

i.e. (8.1) is *elliptic for its solution* φ.

This means that, for a fixed solution φ, we consider δ as given measurable function of the variables (x, y). After this, the equation (8.1) is linear and elliptic.

Let D be a domain in $\mathbb{R}^2$. For a locally Lipschitz function $f : D \to \mathbb{R}$, we denote by $D_b(f)$ the set of all points $a \in D$ in which f is not differentiable. By the Rademacher theorem, the set $D_b(f)$ has zero two-dimensional Lebesgue measure.

We are convenient to use the following definition of generalized solutions of (8.1) [117]. A locally Lipschitz function $\varphi : D \to \mathbb{R}$ is called the *generalized solution* of the equation (8.1), if for every subdomain $\Delta \subset\subset D$ with rectifiable boundary $\partial\Delta$ such that $\text{mes}_1\,(\partial\Delta \cap D_b(\varphi)) = 0$, and for an arbitrary function $\eta \in \text{Lip}\,\overline{\Delta}$, the following relation holds

$$\iint_{\Delta} \delta(q)\,(\varphi_x \eta_x + \varphi_y \eta_y)\,dx\,dy = \int_{\partial\Delta} \eta\,\delta(q)\,(-\varphi_y dx + \varphi_x dy). \qquad (8.3)$$

For the sufficient smooth solution φ, the relation (8.3) implies (8.1) in the traditional sense.

Below we prove the following theorem [124].

Theorem 8.4 *Let φ be a generalized solution of the equation (8.1) in a half-strip $\Pi = \{(x, y) : 0 < x < \infty, 0 < y < \hbar\}$. Assume that for every $0 < x < +\infty$:*

$$\lim_{y \to +0} \varphi(x, y) = \lim_{y \to \hbar - 0} \frac{\partial\varphi}{\partial y}(x, y) = 0, \qquad (8.5)$$

and for some number $s > \pi/\hbar$ for all, sufficiently large $x > 0$:

$$\underset{0 < y < \hbar}{\text{ess sup}}\, |\nabla\varphi(x, y)| \leq \exp\{-\exp\{s\,x\}\}. \qquad (8.6)$$

Then $\varphi \equiv 0$ in Π.

Consider a circular sector of angle α, $0 < \alpha \leq 2\pi$. Namely,

$$D = \{(x, y) \in \mathbb{R}^2 : 0 < \sqrt{x^2 + y^2} < 1, \ y > 0 \text{ and } y\cos\alpha < x\sin\alpha\},$$

if $0 < \alpha \leq \pi$, and
$$D = \{(x,y) \in \mathbb{R}^2 : 0 < \sqrt{x^2+y^2} < 1,\ y > 0 \text{ or } y\cos\alpha < x\sin\alpha\},$$
if $\pi < \alpha \leq 2\pi$.

We put
$$\gamma_0 = \{(x,y) \in \partial D \setminus \{0\} : y = 0\}, \quad \gamma_\alpha = \{(x,y) \in \partial D \setminus \{0\} : \operatorname{arctg}\frac{y}{x} = \alpha\}.$$

Theorem 8.7 *Let φ be a generalized solution of the equation (8.1) in a circular sector D of an angle $0 < \alpha \leq 2\pi$, satisfying boundary conditions*
$$\lim_{(x,y)\to\gamma_\alpha} \varphi(x,y) = \lim_{(x,y)\to\gamma_0} \frac{\partial\varphi}{\partial y}(x,y) = 0. \tag{8.8}$$

Suppose that for some number $s > \pi/\alpha$ and all sufficiently small $r > 0$ it is fulfilled
$$\operatorname*{ess\ sup}_{\sqrt{x^2+y^2}=r} |\nabla\varphi(x,y)| \leq \exp\left\{-\frac{1}{r^s}\right\}. \tag{8.9}$$

Then $\varphi \equiv 0$ in D.

8.2   The Complex Potential

Let D be a simply connected domain. Along with the potential function $\varphi(x,y)$, we consider the stream function $\psi(x,y)$, connected with φ by relations:
$$\begin{cases} \psi_x &= -\delta\varphi_y \\ \psi_y &= \delta\varphi_x \end{cases}. \tag{8.10}$$

The relation (8.3) implies that for every closed rectifiable contour $C \subset D$ with the property $\operatorname{mes}_1 (C \cap D_b(\varphi)) = 0$ it is fulfilled
$$\int_C \delta(q)\ (-\varphi_y dx + \varphi_x dy) = 0.$$

It is clear that for every pair of points $a, b \in D$ a.e. broken lines $C \subset D$, joining a and b, have the showed property. Thus, for a fixed point $a \in D$, we can put
$$\psi(x,y) = \int_a^{(x,y)} \delta(q)\ (-\varphi_y dx + \varphi_x dy), \tag{8.11}$$

and the locally Lipschitz function $\psi(D)$, satisfying (8.10), exists.

The complex valued function $\zeta = \varphi + i\psi$ is called the *complex potential*.

Consider the quadratic form

$$
\begin{aligned}
ds_\Omega^2 &= d\varphi^2 + d\psi^2 = (\varphi_x dx + \varphi_y dy)^2 + (\psi_x dx + \psi_y dy)^2 = \\[2mm]
&= (\varphi_x^2 + \delta^2 \varphi_y^2)\, dx^2 + 2(1 - \delta^2)\, \varphi_x \varphi_y\, dxdy + (\varphi_y^2 + \delta^2 \varphi_x^2)\, dy^2 \equiv \\[2mm]
&\equiv g_{11}\, dx^2 + 2g_{12}\, dxdy + g_{22}\, dy^2.
\end{aligned}
$$

The form is positively defined at every point where $|\nabla\varphi| > 0$.

Consider an abstract surface (D, ds_Ω). Denote by g^{ij} $(i, j = 1, 2)$, coefficients of the matrix (g^{ij}) inverse to (g_{ij}). Put $g = \det(g_{ij})$. We have

$$
g = g_{11}g_{22} - g_{12}^2 = \delta^2 |\nabla\varphi|^4
$$

and further, by relations

$$
g^{11} = \frac{g_{22}}{g}, \quad g^{12} = -\frac{g_{12}}{g}, \quad g^{22} = \frac{g_{11}}{g},
$$

the Laplace-Beltrami equation for functions $u(x, y)$, harmonic with respect to the metric ds_Ω, we rewrite in the form

$$
\left(\sqrt{g}\, g^{12}\, \frac{\partial u}{\partial y} + \sqrt{g}\, g^{11}\, \frac{\partial u}{\partial x} \right)'_x + \left(\sqrt{g}\, g^{12}\, \frac{\partial u}{\partial x} + \sqrt{g}\, g^{22}\, \frac{\partial u}{\partial y} \right)'_y = 0. \qquad (8.12)
$$

Here

$$
g^{11} = \frac{\varphi_y^2 + \delta^2\, \varphi_x^2}{\delta^2\, |\nabla\varphi|^4},
$$

$$
g^{12} = g^{21} = -\frac{(1 - \delta^2)\varphi_x\, \varphi_y}{\delta^2\, |\nabla\varphi|^4}, \qquad (8.13)
$$

$$
g^{22} = \frac{\varphi_x^2 + \delta^2\, \varphi_y^2}{\delta^2\, |\nabla\varphi|^4}.
$$

The Cauchy-Riemann system with respect to the metric ds_Ω is rewritten in the form

$$
\begin{cases}
\dfrac{\partial v}{\partial x} = \dfrac{(1 - \delta^2)\varphi_x\, \varphi_y}{\delta\, |\nabla\varphi|^2}\, \dfrac{\partial u}{\partial x} - \dfrac{\varphi_x^2 + \delta^2\, \varphi_y^2}{\delta\, |\nabla\varphi|^2}\, \dfrac{\partial u}{\partial y}, \\[4mm]
\dfrac{\partial v}{\partial y} = \dfrac{\varphi_y^2 + \delta^2\, \varphi_x^2}{\delta\, |\nabla\varphi|^2}\, \dfrac{\partial u}{\partial x} - \dfrac{(1 - \delta^2)\varphi_x\, \varphi_y}{\delta\, |\nabla\varphi|^2}\, \dfrac{\partial u}{\partial y}
\end{cases}
\qquad (8.14)
$$

(see, for example, [163, Chapter 1, §1]).

Now let's check the following statement.

Lemma 8.15 *The function $\zeta(x,y)$ is holomorphic with respect to the metric ds_Ω.*

For the **proof** it is sufficient to remark that for $u = \varphi$ and $v = \psi$ the system (8.14) transforms into the system (8.10).

8.3 Maps onto Strips

Let D be a simply connected domain different from $\mathbb{R}^2$, and let φ be a generalized solution of the equation (8.1). Consider a family of closed locally rectifiable arcs $\gamma \in \Gamma$ lying in D and such that $\mathrm{mes}_1(\gamma \cap D_b(\varphi)) = 0$. A Lebesgue measurable, locally bounded nonnegative function $\rho(x,y) : D \to \mathbb{R}$ is admissible for the family Γ with respect to the metric ds_Ω if for every arc $\gamma \in \Gamma$:

$$\int_\gamma \rho(x,y)\, ds_\Omega \geq 1. \tag{8.16}$$

We define a module of Γ in the metric ds_Ω setting

$$\mathrm{mod}_\Omega \Gamma = \inf \iint_D \rho^2(x,y)\, d\sigma_\Omega \tag{8.17}$$

where the infimum is taken over all functions $\rho(x,y)$ admissible for Γ with respect to the metric of the surface (D, ds_Ω).

Fix three prime ends e', e_0 and e'' on the boundary of D lying in the order of the positive direction of the circuit of the boundary $\tilde{D} \setminus D$. Suppose that at a neighborhood of e' the surface (D, ds_Ω) satisfies (7.18). Suppose also that

$$0 < \underset{A}{\mathrm{ess\,inf}}\, \delta(|\nabla\varphi(x,y)|) \leq \underset{A}{\mathrm{ess\,sup}}\, \delta(|\nabla\varphi(x,y)|) < \infty \tag{8.18}$$

for every compact $A \subset \tilde{D} \setminus \{e'\}$.

It is clear that (8.18) implies (7.18) for every prime end $e \in \tilde{D} \setminus \{e'\}$.

Let $f : D \to \mathbb{R}^2$ be a schlicht conformal (in the metric of (D, ds_Ω)) map with properties

$$f(D) = \Pi, \quad \Pi = \{(u,v) \in \mathbb{R}^2 : -\frac{\pi}{2} < v < \frac{\pi}{2}\},$$

and

$$\lim_{(x,y)\to e'} u(x,y) = +\infty, \quad f(e_0) = (0,1), \quad \lim_{(x,y)\to e''} u(x,y) = -\infty. \qquad (8.19)$$

We fix a locally Lipschitz function $h : D \to [0,1]$ for which

$$\lim_{(x,y)\to e'} h(x,y) = 0, \quad \lim_{(x,y)\to \widetilde{e_0 e''}} h(x,y) = 1. \qquad (8.20)$$

Suppose that h satisfies the property (7.16) too.

For an arbitrary arc $E_h(t)$, $0 < t < 1$, by $\mathcal{E}(t)$ we denote the family of all locally rectifiable arcs γ, $\mathrm{mes}_1 (\gamma \cap D_b(\varphi)) = 0$, such that $[\gamma]_{\widetilde{D}}$ separates on $\widetilde{D}$ the end e' and the arc $\widetilde{e_0 e''}$ with $[\gamma]_{\widetilde{D}} \cap [E_h(t)]_{\widetilde{D}} \neq \emptyset$. Let $\Lambda(t)$ be the family of arcs separating $E_h(t)$ and $\widetilde{e_0 e''}$ on D.

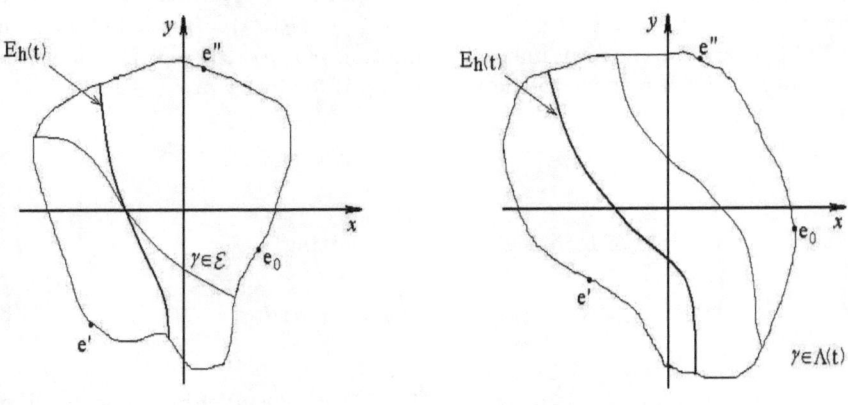

Fig. 8.1.

The following statement is a simple corollary of the Ahlfors-Warshawski type theorem.

Lemma 8.21 *Under described assumptions for a schlicht and conformal, with respect to (D, ds_Ω), map f with the normalization (8.19), it is fulfilled*

$$u(x,y) \leq \pi \left(\mathrm{mod}_\Omega \Lambda(t) + \mathrm{mod}_\Omega \mathcal{E}(t) \right) \quad \text{for all} \quad (x,y) \in E_h(t) \qquad (8.22)$$

and

$$\mathrm{osc}\,(u(x,y), E_h(t)) \leq 2\pi \, \mathrm{mod}_\Omega \mathcal{E}(t). \qquad (8.23)$$

For the **proof** it is sufficient to turn to Theorem 7.63.   □

8.4   The Uniqueness Problem

We define a class of holomorphic functions of the *bounded kind A* as the class
of the functions, regular in $|z| < R$ and satisfying the condition

$$\sup_{\rho < R} \int_0^{2\pi} \ln^+ |f(\rho e^{i\varphi})|\, d\varphi < \infty \quad (\ln^+ x = \max\{\ln x, 0\}).$$

Moreover, we put

$$A(f, \rho) = \frac{1}{2\pi} \int_0^{2\pi} \ln^+ |f(\rho e^{i\varphi})|\, d\varphi, \quad A(f) = \sup_{\rho < R} A(f, \rho).$$

Lemma 8.24   *If a function $f(z)$ is regular in a ring $r_1 < |z| < r_2$, then
functions*

$$A(f, z) = \frac{1}{2\pi} \int_0^{2\pi} \ln^+ |f(z\, e^{i\varphi})|\, d\varphi,$$

$$L(f, z) = \frac{1}{2\pi} \int_0^{2\pi} \ln |f(z\, e^{i\varphi})|\, d\varphi$$

are subharmonic in $r_1 < |z| < r_2$ and depend only on $|z|$.

Proof. It is sufficient to prove the statement for $A(f, z)$. Considerations
for the second case are analogous.

At first we remark that the function $A(f, |z|e^{i\theta})$ is independent of θ, because
an integral of a periodical function with respect to its period does not change
under changes of the beginning of the integration segment.

Further, the integral for $A(f, z)$ is a limit of integral sums

$$S_n(z) = \frac{1}{2\pi} \sum \ln^+ |f(ze^{i\varphi_k})|\, (\varphi_{k+1} - \varphi_k);$$

moreover, the convergence to the limit is uniform with respect to z on every
subring lying inside of the original ring. Every function $S_n(z)$ is subharmonic

as a sum of subharmonic functions. The limit of these sums $A(f,z)$ is also subharmonic.   □

We derive the following statement from the maximum principle for subharmonic functions.

Corollary 8.25   *If $f(z)$ is regular in $|z| < R$, then functions $A(f,z)$ and $L(f,z)$ are nondecreasing functions of the the variable $\rho = |z|$.*

The theorem formulated below states that the functions of the boun- ded kind can not tend to zero arbitrarily rapidly as $|z| \to R$.

Theorem 8.26   *If $f(z)$ is a function of the bound kind in $|z| < R$ and if*

$$\lim_{\rho \to R} \int_0^{2\pi} \ln |f(\rho\, e^{i\varphi})|\, d\varphi = -\infty,$$

then $f(z) \equiv 0$.

Proof. Since $\ln^+ x \geq \ln x$, then from $f \in A$ the upper boundedness of the integral $A(f,\rho)$ follows. Corollary 8.25 guarantees that this integral is the nondecreasing function of ρ. Thus the assumption of the theorem can be valid if and only if the integral does not depend on ρ and identically equals to $-\infty$. It is possible only in the case $f(z) \equiv 0$.   □

This theorem is the generalization of the classical uniqueness statement:

If a function $f(z)$ is regular in a neighborhood of a point $z = a$ and, as $z \to a$, the function $f(z)$ tends to zero more rapid than an arbitrary power of $z - a$, then $f(z) \equiv 0$.

The following statement is also a generalization of the uniqueness theorem.

Theorem 8.27   *Let $f(z)$ be a function of the bounded kind in $|z| < R$ with zero at points $z_1, z_2, \ldots$[1]. If*

$$\sum_{n=1}^{\infty} (R - |z_n|) = +\infty,$$

then $f(z) \equiv 0$.

---

[1] A multiple zero is written according to its multiplicity.

Proof. Suppose that $f(z) \not\equiv 0$. Without loss of generality, we can assume that $f(0) \neq 0$. It is clear, because in the opposite case instead of $f(z)$ we can consider the function $g(z) = z^{-m} f(z)$ where m is the multiplicity of zero of $f(z)$ at the point $z = 0$. The function $g(z)$ has also the bounded kind.

We use the well-known Jensen formula

$$\frac{1}{2\pi} \int_0^{2\pi} \ln |f(\rho e^{i\varphi})| \, d\varphi = \ln |f(0)| - \sum_{|z_n| < \rho} \ln \frac{\rho}{|z_n|}$$

which is valid for an arbitrary regular function $f(z)$ in $|z| \leq R$, with zeros at $z_1, z_2, \ldots, z_n, \ldots$ (see, for example, [35, Chapter VIII, §2]).

Passing to limit as $\rho \to R$, we have

$$\lim_{\rho \to R} \frac{1}{2\pi} \int_0^{2\pi} \ln |f(\rho e^{i\varphi})| \, d\varphi = \ln |f(0)| - \sum_{n=1}^{\infty} \ln \frac{\rho}{|z_n|}.$$

However, for $0 < |z| < R$

$$\ln \frac{R}{|z|} = -\ln \left(1 - \frac{R - |z|}{R} \right) = \sum_{n=1}^{\infty} \frac{1}{n} \left(\frac{R - |z|}{R} \right)^n > \frac{R - |z|}{R}$$

and, therefore,

$$\sum_{n=1}^{\infty} \ln \frac{R}{|z_n|} \geq \sum_{n=1}^{\infty} \frac{R - |z_n|}{R} = +\infty.$$

Thus

$$\lim_{\rho \to R} \int_0^{2\pi} \ln |f(\rho e^{i\varphi})| \, d\varphi = -\infty,$$

and by Theorem 8.26 we obtain $f(z) \equiv 0$. □

For arbitrary domains of the plane, this theorem can be rewritten, for example, in the following form.

Theorem 8.28 *Let D be a simply connected plane domain and let $w(z)$ be a conformal map of D onto $|w| < 1$. If $f(z)$ is regular, bounded in D, has zeros at points $z_1, z_2, \ldots$, and*

$$\sum_{n=1}^{\infty} (1 - |w(z_n)|) = +\infty,$$

then $f(z) \equiv 0$.

Proof. Consider the function

$$g(w) = f(z(w))$$

where $z(w)$ is a map inverse to $w(z)$. The function $g(w)$ is regular and bounded in $|w| < 1$, has zeros at points $w_n = w(z_n)$. Moreover,

$$\sum_{n=1}^{\infty}(1 - |w_n|) = \sum_{n=1}^{\infty}(1 - |w(z_n)|) = +\infty.$$

By Theorem 8.27, we have $g(w) \equiv 0$. □

Now we will show functions which illustrate the exactness of Theorems 8.26 and 8.27. Let $a(\varphi)$ be a continuous function such that

$$\int_0^{2\pi} \ln a(\varphi)\, d\varphi > -\infty.$$

It is not difficult to check that the function

$$f(z) = \frac{1}{2\pi}\int_0^{2\pi} \frac{R\,e^{i\varphi} + z}{R\,e^{i\varphi} - z}\ln a(\varphi)\, d\varphi \quad (|z| < R)$$

is holomorphic and bounded in $|z| < R$. Moreover,

$$|f(z)| \to a(\varphi) \quad \text{for} \quad z \to R\,e^{i\varphi}.$$

Another example. Suppose that there is a sequence $\{z_n\}$ such that

$$|z_n| < R, \quad \sum_{n=1}^{\infty}(R - |z_n|) < \infty.$$

The function

$$f(z) = \prod_{n=1}^{\infty} \frac{R(z_n - z)}{R^2 - z\bar{z}_n}\, e^{-i\varphi_n} \quad (\varphi_n = \arg z_n)$$

is holomorphic, $|f(z)| \le 1$ for $|z| < R$ and $f(z_n) = 0$.

We need to remark that the most difficulty under applications of Theorem 8.28 supply estimates of $w(z_n)$ as $n \to \infty$. Proved before in Section 7.1, Ahlfors and Warschawski theorems can be used with a view.

We formulate Theorem 8.26 in the more simple form:

Suppose that $f(z)$ is holomorphic in $|z| < 1$ and continuous up to the boundary. If

$$\int_0^{2\pi} \ln |f(e^{i\varphi})| \, d\varphi = -\infty,$$

then $f(z) \equiv 0$.

With an auxiliary conformal map, this statement can be easily extended for other domains. At first we prove a corresponding result for the half-plane.

Theorem 8.29   *Assume that a function $f(z)$ is regular in $\operatorname{Re} z > 0$ and continuous in $\operatorname{Re} z \geq 0$. Assume also that there exists a continuous positive function $\nu(t) : \mathbb{R}_+ \to \mathbb{R}$ for which*

$$\ln |f(z)| < -\nu(|z|) \quad (\operatorname{Re} z \geq 0).$$

If

$$\int_1^\infty \frac{\nu(t)}{t^2} = +\infty,$$

then $f(z) \equiv 0$.

Proof. The function

$$z = \frac{1+w}{1-w}$$

maps conformally the disc $|w| < 1$ onto the half-plane $\operatorname{Re} z > 0$. Thus the function

$$F(w) = f\left(\frac{1+w}{1-w}\right)$$

is holomorphic in $|w| < 1$ and continuous up to the boundary. We have

$$\int_0^{2\pi} \ln |F(e^{i\varphi})| \, d\varphi = \int_{-\pi}^{\pi} \ln \left| \frac{1+e^{i\varphi}}{1-e^{i\varphi}} \right| \, d\varphi < -2 \int_0^\pi \nu\left(\operatorname{ctg}\frac{\varphi}{2}\right) d\varphi.$$

However,

$$\int_0^\pi \nu\left(\operatorname{ctg}\frac{\varphi}{2}\right) d\varphi = 2 \int_0^{\pi/2} \nu(\operatorname{ctg}\theta) \, d\theta =$$

$$= 2 \int_0^\infty \frac{\nu(t)dt}{1+t^2} \geq 2 \int_1^\infty \frac{\nu(t)dt}{1+t^2} \geq \int_1^\infty \frac{\nu(t)}{t^2} dt = +\infty.$$

By Theorem 8.26 we have $F(w) \equiv 0$, and hence, $f(z) \equiv 0$. $\square$

Analogously, we prove:

Theorem 8.30   *Let $f(z)$ be a function regular in $|\operatorname{Im} z| < \frac{\pi}{2}$ and continuous in its closure. Suppose that it satisfies the inequality*

$$\ln |f(x + iy)| < -\nu(x) \quad \left(-\frac{\pi}{2} < y < \frac{\pi}{2}\right)$$

where $\nu(t)$ is a continuous positive function. If

$$\int\limits_0^\infty \nu(t)\, e^{-t} dt = +\infty,$$

then $f(z) \equiv 0$.

Using the Warschawski theorem, let's prove the analogous statement for strip regions of the sufficient form. Namely, let D be a domain, given by inequalities

$$\varphi(x) - \frac{1}{2}\theta(x) < y < \varphi(x) + \frac{1}{2}\theta(x) \quad (-\infty < x < \infty)$$

where $\varphi(x)$ and $\theta(x)$ are continuously differentiable functions with properties

$$|\varphi'(x)| < M, \quad |\theta'(x)| < M, \quad \int\limits_0^\infty \frac{\theta'^2(x)}{\theta(x)}\, dx < +\infty.$$

Theorem 8.31   *Suppose that a function $f(z)$ is holomorphic in the domain D, continuous in $\overline{D}$, and satisfies the assumption*

$$\ln |f(x + iy)| < -\nu(x) \quad (x + iy \in D)$$

where $\nu(x)$ is a continuous positive nondecreasing function. Let

$$s(x) = \pi \int\limits_0^x \frac{1 + \varphi'^2(t)}{\theta(t)}\, dt.$$

If

$$\int\limits_0^\infty \nu(x) e^{-s(x)} \frac{dx}{\theta(x)} = +\infty,$$

then $f(z) \equiv 0$.

Proof. Denote the conformal map of D onto $|\operatorname{Im} w| < \frac{\pi}{2}$ by $w(x)$ such that

$$\operatorname{Re} w \to \pm\infty \quad \text{for} \quad \operatorname{Re} z \to \pm\infty.$$

Under these assumptions we may to use the Warschawski Theorem 7.2. We have

$$\operatorname{Re} w(x + iy) - \operatorname{Re} w(a + ib) < s(x) - s(a) + C' \quad (x > a)$$

where the constant C' does not depend on x, y, a, b. Setting

$$a = 0, \quad C = C' + \sup_b \operatorname{Re} w(ib),$$

we arrive to the inequality

$$\operatorname{Re} w(x + iy) < s(x) + C' \quad (x > 0). \tag{8.32}$$

Denote by $z(w)$ the map inverse to $w(z)$. Further,

$$x(u) = \min \operatorname{Re} z(w) \quad \left(\operatorname{Re} w = u, \ |\operatorname{Im} w| \le \frac{\pi}{2}\right).$$

Choosing w in (8.32) as a value, for which

$$\operatorname{Re} w = u \quad \operatorname{Re} z(w) = x(u),$$

we obtain

$$u < s(x(u)) + C. \tag{8.33}$$

Since $\nu(x)$ is nondecreasing, the function

$$F(w) = f(z(w)) \quad \left(|\operatorname{Im} w| < \frac{\pi}{2}\right)$$

is holomorphic, continuous in the closed strip, and satisfies the inequality

$$\ln|F(u + iv)| = \ln|f(z(u + iv))| < -\nu(x(u)).$$

By the inequality (8.33), we deduce

$$x(u) < k(u - C)$$

where $k(u)$ is a function inverse to $s(x)$. Moreover, it is clear that the function $k(u)$ does not decrease and that $k(u) \to +\infty$ as $u \to +\infty$. From this we find

$$\ln|F(u + iv)| < -\nu(k(u - C))$$

and

$$e^C \int_C^\infty \nu(k(u-C))e^{-u}du =$$

$$= \int_0^\infty \nu(x)e^{-s(x)}\, s'(x)\, dx > \int_0^\infty \nu(x)\, e^{-s(x)}\frac{dx}{\theta(x)} = +\infty\,.$$

By Theorem 8.30 we conclude that $F(w) \equiv 0$. This means that $f(z) \equiv 0$, that is needed. □

8.5   Phragmén-Lindelöf Theorems

We will need the following version of the Phragmén-Lindelöf theorem.

Theorem 8.34   *Let* $F : \Pi \to \mathbb{R}^2$ *be a holomorphic in* Π *and satisfying to the inequality*

$$\ln |F(u,v)| < -\nu(u) \quad \left(-\frac{\pi}{2} < v < \frac{\pi}{2}\right) \tag{8.35}$$

where $\nu(u)$ *is a positive, nondecreasing, continuous function. If*

$$\int^{+\infty} \nu(u)\, e^{-u}du = +\infty, \tag{8.36}$$

then $F(u,v) \equiv 0$.

For the **proof** we remark that in the case of the Euclidean metric and assumption to the continuity of F up to the boundary, this statement is contained in Theorem 8.30. In the general case, the proof is practically the same. It is sufficient to consider F in a more narrow strip $\{(u,v) \in \mathbb{R}^2 : |v| < c < \frac{\pi}{2}\}$, $c \equiv$ const, and to pass to the limit. □

The following Phragmén-Lindelöf type theorem for holomorphic functions with respect to a metric ds_Ω has a preparatory character.

Theorem 8.37   *Let* $D \subset \mathbb{R}^2$ *be a simply connected domain, different from* $\mathbb{R}^2$. *Let* e', e_0. *and* e'' *be prime ends in* $\partial\tilde{D}$ *and* $h : D \to \mathbb{R}$ *be a locally Lipschitz function with* (8.20), (7.16). *Suppose that a solution* φ *of* (8.1) *satisfies* *conditions* (8.18), (7.18). *Let* $\Phi : D \to \mathbb{R}^2$ *be a function which is holomorphic with respect to the metric* ds_Ω *and such that*

$$\ln |\Phi(x,y)| \leq -\nu(t) \quad \text{for all} \quad (x,y) \in E_h(t) \quad (0 < t < 1) \tag{8.38}$$

for a positive nondecreasing continuous function ν.

Let $\tau = \theta(t)$ *be a strictly monotone decreasing function, continuous in* $(0,1)$, $\theta(+0) = +\infty$ *for which*

$$\pi \left(\text{mod}_\Omega \Lambda(t) + \text{mod}_\Omega \mathcal{E}(t) \right) \leq \theta(t) \quad (0 < t < 1). \tag{8.39}$$

If

$$\int\limits^{+\infty} \nu \left(\theta^{-1}(\tau) \right) e^{-\tau} \, d\tau = +\infty, \tag{8.40}$$

then $\Phi(x,y) \equiv 0$.

Proof. Denote by f the schlicht map, conformal with respect to ds_Ω, transforming the quadratic form ds_Ω^2 to the form $\lambda(u,v) \, (du^2 + dv^2)$. By assumptions (7.18), (8.18), we may assume that $f(D) = \Pi$ and (8.19) is satisfied.

We put $F(u,v) = \Phi \circ f^{-1}(u,v)$. This function is holomorphic in the strip Π with respect to the Euclidean metric. The condition (8.38) implies

$$\sup_{D_t} |\Phi(x,y)| \leq e^{-\nu(t)}$$

where

$$D_t = \{(x,y) \in D : h(x,y) \leq t\}.$$

Thus

$$\sup_{f(D_t)} |\Phi \circ f^{-1}(u,v)| \leq e^{-\nu(t)}.$$

However, by (8.22) and (8.39), it is fulfilled

$$u \leq \theta(t) \quad \text{for all} \quad (u,v) \in D \setminus f(D_t).$$

The function $\nu(t)$ is nondecreasing, and hence for all $(\tau, v) \in \Pi$, we have

$$|\Phi \circ f^{-1}(\tau, v)| \leq e^{-\nu(\theta^{-1}(\tau))}$$

and, further,

$$\ln |F(\tau, v)| \leq -\nu_1(\tau), \qquad \nu_1(\tau) = \nu \left(\theta^{-1}(\tau) \right).$$

The condition (8.40) implies

$$\int\limits^{+\infty} \nu_1(\tau) e^{-\tau} = \infty.$$

By Theorem 8.34 we conclude that $F(u,v) \equiv 0$. Thus $\Phi(x,y) \equiv 0$, and the theorem is proved. $\qquad \square$

8.6   Two Lemmas

Fix t, $0 < t < 1$, and a subdomain D_t, defined in the previous section. We estimate $\mathrm{mod}_\Omega \Lambda(t)$. Consider a family $\Lambda^*(t)$ of all locally rectifiable arcs γ lying in $D \setminus D_t$, satisfying the condition

$$\mathrm{mes}\,(\gamma \cap D_b(\varphi)) = 0\,,$$

and joining in $D \setminus D_t$ the boundary arc $\widetilde{e_0 e''}$ with $E_h(t)$. By (1.26) we have

$$\mathrm{mod}_\Omega \Lambda(t) = \frac{1}{\mathrm{mod}_\Omega \Lambda^*(t)}. \tag{8.41}$$

Further, we use the well-known connection between the module and the capacity of a condenser 1.43. We have

$$\mathrm{mod}_\Omega \Lambda^*(t) = \mathrm{cap}_\Omega \left(E_h(t), \widetilde{e_0 e''}; D \setminus D_t \right) \tag{8.42}$$

where

$$\mathrm{cap}_\Omega(E_h(t), \widetilde{e_0 e''}; D \setminus D_t) = \inf \iint_{D \setminus D_t} (g^{11}\xi_x^2 + 2g^{12}\xi_x\xi_y + g^{22}\xi_y^2)\, d\sigma_\Omega$$

and the infimum is taken over all locally Lipschitz functions $\xi : D \setminus D_t \to \mathbb{R}$ for which

$$\xi \big|_{\widetilde{e_0 e''}} = 1, \quad \xi \big|_{E_h(t)} = 0\,.$$

It is easy to see that

$$\iint_{D \setminus D_t} (g^{11}\xi_x^2 + 2g^{12}\xi_x\xi_y + g^{22}\xi_y^2)\, d\sigma_\Omega =$$

$$\iint_{D \setminus D_t} \left(\frac{\varphi_y^2 + \delta^2 \varphi_x^2}{\delta |\nabla \phi|^2} \xi_x^2 - 2 \frac{1 - \delta^2}{\delta |\nabla \varphi|^2} \varphi_x \varphi_y \xi_x \xi_y + \frac{\varphi_x^2 + \delta^2 \varphi_y^2}{\delta |\nabla \phi|^2} \xi_y^2 \right) dx dy =$$

$$\iint_{D \setminus D_t} \left\{ \frac{\varphi_y^2 \xi_x^2 - 2\varphi_x \varphi_y \xi_x \xi_y + \varphi_x^2 \xi_y^2}{\delta |\nabla \varphi|^2} + \delta \frac{\varphi_x^2 \xi_x^2 + 2\varphi_x \varphi_y \xi_x \xi_y + \varphi_y^2 \xi_y^2}{|\nabla \varphi|^2} \right\} dx dy =$$

$$\iint_{D \setminus D_t} \left\{ \delta \left\langle \nabla \xi, \frac{\nabla \varphi}{|\nabla \varphi|} \right\rangle^2 + \frac{1}{\delta} \left\langle \nabla \xi, \frac{(\nabla \varphi)^\perp}{|\nabla \varphi|} \right\rangle^2 \right\} dx dy$$

where $\langle \cdot, \cdot \rangle$ means the scalar product and $(\cdot)^\perp$ - the orthogonal complement of a vector.

Observing now that

$$\delta \left\langle \nabla \xi, \frac{\nabla \varphi}{|\nabla \varphi|} \right\rangle^2 + \frac{1}{\delta} \left\langle \nabla \xi, \frac{(\nabla \varphi)^\perp}{|\nabla \varphi|} \right\rangle^2 \geq$$

$$\geq \min \left(\delta, \frac{1}{\delta} \right) \left\{ \left\langle \nabla \xi, \frac{\nabla \varphi}{|\nabla \varphi|} \right\rangle^2 + \left\langle \nabla \xi, \frac{(\nabla \varphi)^\perp}{|\nabla \varphi|} \right\rangle^2 \right\} =$$

$$= \min \left(\delta, \frac{1}{\delta} \right) |\nabla \xi|^2,$$

we obtain

$$\iint\limits_{D \backslash D_t} (g^{11} \xi_x^2 + 2 g^{12} \xi_x \xi_y + g^{22} \xi_y^2)\, d\sigma_\Omega \geq \iint\limits_{D \backslash D_t} \min \left(\delta, \frac{1}{\delta} \right) |\nabla \xi|^2\, dx dy.$$

Taking into account (8.41) and (8.42), we arrive to the statement:

Lemma 8.43 *The following estimate holds*

$$\mathrm{mod}_\Omega \Lambda(t) \leq 1 \bigg/ \inf_\xi \iint_{D \backslash D_t} \min \left(\delta, \frac{1}{\delta} \right) |\nabla \xi|^2\, dx dy \, . \qquad (8.44)$$

Now we will show another useful estimate for the module of an arc family with respect to the metric ds_Ω. For a family Γ of locally rectifiable arcs γ, we have

$$\mathrm{mod}_\Omega \Gamma = \inf_{\rho \geq 0} \frac{\iint_D \rho^2(x, y) \delta |\nabla \varphi|^2 dx dy}{\left(\inf_\gamma \int\limits_\gamma \rho(x, y)\, ds_\Omega \right)^2} =$$

$$= \inf_{\widetilde{\rho} \geq 0} \frac{\iint_D \widetilde{\rho}^2(x, y) \delta\, dx dy}{\left(\inf_\gamma \int\limits_\gamma \widetilde{\rho}(x, y)\, d\widetilde{s}_\Omega \right)^2}$$

where $\widetilde{\rho} = \rho \, |\nabla\varphi|$ and

$$d\widetilde{s}_\Omega^2 = \frac{\varphi_x^2 + \delta^2\varphi_y^2}{|\nabla\varphi|^2}dx^2 + 2(1 - \delta^2)\frac{\varphi_x\varphi_y}{|\nabla\varphi|^2}dxdy + \frac{\varphi_y^2 + \delta^2\varphi_x^2}{|\nabla\varphi|^2}dy^2.$$

It is easy to see that

$$d\widetilde{s}_\Omega^2 = \frac{(\varphi_x dx + \varphi_y dy)^2}{|\nabla\varphi|^2} + \frac{(\varphi_y dx - \varphi_x dy)^2}{|\nabla\varphi|^2} \geq$$

$$\geq \min(1, \delta)(dx^2 + dy^2).$$

Therefore, setting

$$\widetilde{\rho}(x, y) = \rho(x, y) \, \min(1, \delta)$$

and observing that

$$\frac{\delta}{\min(1, \delta^2)} = \max\left(\delta, \frac{1}{\delta}\right),$$

we arrive to the relation

$$\frac{\displaystyle\iint_D \rho^2(x, y) \, d\sigma_\Omega}{\left(\displaystyle\inf_\gamma \int_\gamma \rho(x, y) \, ds_\Omega\right)^2} \leq \frac{\displaystyle\iint_D \widetilde{\rho}^2(x, y) \, \max(\delta, \frac{1}{\delta}) \, dxdy}{\left(\displaystyle\inf_\gamma \int_\gamma \widetilde{\rho}(x, y) \, \sqrt{dx^2 + dy^2}\right)^2}.$$

Thus the following statement is proved:

Lemma 8.45   *Under described assumptions, the following estimate holds*

$$\mathrm{mod}_\Omega \Gamma \leq \inf_{\rho \geq 0} \frac{\displaystyle\iint_D \rho^2(x, y) \max\left(\delta, \frac{1}{\delta}\right) dxdy}{\left(\displaystyle\inf_\gamma \int_{\gamma \in \Gamma} \rho(x, y) \, \sqrt{dx^2 + dy^2}\right)^2}. \qquad (8.46)$$

8.7 Circular Sectors

We consider a circular sector of an angle α, $0 < \alpha \le 2\pi$. Fix prime ends

$$e' = (0,0), \quad e_0 = (1,0), \quad e'' = (\cos\alpha, \sin\alpha),$$

and the function

$$h(x,y) = \sqrt{x^2 + y^2}.$$

It is easy to see that h satisfies (7.16) and (8.20).

For an arbitrary $0 < r < 1$, we put

$$S_D(0,r) = D \cap \{(x,y) : x^2 + y^2 = r^2\},$$

$$B_D(0,r) = D \cap \{(x,y) : x^2 + y^2 < r^2\}.$$

In applications of Theorem 8.37, the greatest difficulty is to obtain a suitable estimate for $\mathrm{mod}_\Omega \mathcal{E}(t)$. Below we bring one of such estimates. Fix t, $0 < 2t < 1$, and denote by $\mathcal{E}_1(t)$ the set of all arcs $\gamma \in \mathcal{E}(t)$ lying in $B_D(0, 2t) \setminus B_D(0, t/2)$, by $\mathcal{E}_2(t)$ - the set of arcs $\gamma \in \mathcal{E}(t) \setminus \mathcal{E}_1(t)$. We have

$$\mathrm{mod}_\Omega \mathcal{E}(t) \le \mathrm{mod}_\Omega \mathcal{E}_1(t) + \mathrm{mod}_\Omega \mathcal{E}_2(t). \tag{8.47}$$

We use Lemma 8.45. By (8.46) we have

$$\mathrm{mod}_\Omega \mathcal{E}_1(t) \le \inf_{\rho \ge 0} \frac{\iint_D \rho^2(x,y) \max\left(\delta, \frac{1}{\delta}\right) dx dy}{\left(\inf_\gamma \int_{\gamma \in \mathcal{E}_1(t)} \rho(x,y) \sqrt{dx^2 + dy^2}\right)^2}.$$

Setting here $B_D^*(t) = B_D(0, 2t) \setminus B_D(0, t/2)$ and

$$\rho(x,y) = \begin{cases} \dfrac{1}{h(x,y)} & \text{for} \quad (x,y) \in B_D^*(t), \\ 0 & \text{outside of} \quad B_D^*(t), \end{cases}$$

we obtain

$$\mathrm{mod}_\Omega \mathcal{E}_1(t) \le \frac{\iint_{B_D^*(t)} \max(\delta, 1/\delta) \, h^{-2}(x,y) \, dx dy}{\left(\inf_\gamma \int_{\gamma \in \mathcal{E}_1(t)} h^{-1}(x,y) \sqrt{dx^2 + dy^2}\right)^2}.$$

Since for every arc $\gamma \in \mathcal{E}_1(t)$, it is fulfilled

$$\int\limits_{\gamma \in \mathcal{E}_1(t)} h^{-1}(x,y)\sqrt{dx^2 + dy^2} \geq \alpha,$$

we arrive to the estimate

$$\mathrm{mod}_\Omega \mathcal{E}_1(t) \leq \frac{1}{\alpha^2} \iint\limits_{B_D^*(t)} \max(\delta, 1/\delta)\, \frac{dxdy}{h^2(x,y)}. \tag{8.48}$$

To estimate $\mathrm{mod}_\Omega \mathcal{E}_2(t)$, we remark at first that every arc $\gamma \in \mathcal{E}_2(t)$ intersects simultaneously $S_D(0,t)$ and at least one of arcs $S_D(0,t/2)$, $S_D(0,2t)$. From this we conclude that

$$\mathrm{mod}_\Omega \mathcal{E}_2(t) \leq \mathrm{mod}_\Omega \mathcal{E}_3(t) + \mathrm{mod}_\Omega \mathcal{E}_4(t) \tag{8.49}$$

where

$$\mathcal{E}_3(t) = \{\gamma \in \mathcal{E}(t) : \gamma \cap S_D(0,t/2) \neq \emptyset\}$$

and

$$\mathcal{E}_4(t) = \{\gamma \in \mathcal{E}(t) : \gamma \cap S_D(0,2t) \neq \emptyset\}.$$

In every arc $\gamma \in \mathcal{E}_3(t)$, we choose a subarc γ^*, lying in the domain $D \cap \{t/2 < h(x,y) < t\}$ and joining $S_D(0,t/2)$, $S_D(0,t)$. Let $\mathcal{E}_3^*(t)$ be the family of all such arcs γ^*. It is easy to see that

$$\mathrm{mod}_\Omega \mathcal{E}_3(t) \leq \mathrm{mod}_\Omega \mathcal{E}_3^*(t).$$

Choosing now in (8.46) the density $\rho(x,y) = 1/h(x,y)$ for $(x,y) \in B_D^{**}(t)$ where $B_D^{**}(t) = B_D(0,t) \setminus B_D(0,t/2)$, and defining it by zero in all other points, we obtain

$$\mathrm{mod}_\Omega \mathcal{E}_3^*(t) \leq \frac{\displaystyle\iint\limits_{B_D^{**}(t)} \max(\delta, 1/\delta)\, h^{-2}(x,y)\, dxdy}{\left(\displaystyle\inf_\gamma \int\limits_{\gamma^* \in \mathcal{E}_3^*(t)} h^{-1}(x,y)\sqrt{dx^2 + dy^2}\right)^2} \leq$$

$$\leq \ln^{-2} 2 \iint\limits_{B_D^{**}(t)} \max(\delta, 1/\delta)\, \frac{dxdy}{h^2(x,y)}.$$

Thus

$$\mathrm{mod}_\Omega \mathcal{E}_3(t) \leq \ln^{-2} 2 \iint\limits_{B_D^{**}(t)} \max(\delta, 1/\delta)\, \frac{dxdy}{h^2(x,y)}. \tag{8.50}$$

Analogously,

$$\text{mod}_\Omega \mathcal{E}_4(t) \leq \ln^{-2} 2 \iint_{B_D^{**}(t)} \max\left(\delta, 1/\delta\right) \frac{dxdy}{h^2(x,y)}. \tag{8.51}$$

Combining (8.47)–(8.51), we arrive to the following statement:

Lemma 8.52 *If $0 < 2t < 1$, then*

$$\text{mod}_\Omega \mathcal{E}(t) \leq \left(\frac{2}{\ln^2 2} + \frac{1}{\alpha^2}\right) \iint_{B_D^{**}(t)} \max\left(\delta, \frac{1}{\delta}\right) \frac{dxdy}{x^2 + y^2}. \tag{8.53}$$

Estimate $\text{mod}_\Omega \Lambda(t)$. Here we use Lemma 8.43. We put

$$\mu_*(t) = \inf_{D \setminus B_D(0,t)} \min\left(\delta, 1/\delta\right).$$

We have

$$\inf_\xi \iint_{D \setminus B_D(0,t)} \min\left(\delta, \frac{1}{\delta}\right) |\nabla \xi|^2 \, dxdy \geq$$

$$\geq \mu_*(t) \inf_\xi \iint_{D \setminus B_D(0,t)} |\nabla \xi|^2 \, dxdy = \mu_*(t) \, \text{cap}(S_D(0,t), \widetilde{e_0 e''}; D \setminus B_D(0,t))$$

where by $\text{cap}(A, B; C)$ we denote the standard conformal capacity (1.41) of the condenser $(A, B; C)$.

Because D is the corner domain of the angle $\alpha > 0$, then

$$\text{cap}(S_D(0,t), \widetilde{e_0 e''}; D \setminus B_D(0,t)) = \frac{\alpha}{\ln 1/t}.$$

Thus we arrive to the estimate

$$\inf_\xi \iint_{D \setminus B_D(0,t)} \min\left(\delta, \frac{1}{\delta}\right) |\nabla \xi|^2 \, dxdy \geq \frac{\alpha \mu_*(t)}{\ln 1/t}. \tag{8.54}$$

Theorem 8.55 *Let D be a circular sector of the angle $0 < \alpha \leq 2\pi$ with the vertex at the origin. Let φ be a generalized solution of (8.1) satisfying to (8.18) and such that*

$$\int_0 dt \left/ \int_{S_D(0,t)} (\delta + \delta^{-1}) \sqrt{dx^2 + dy^2} = +\infty. \right. \tag{8.56}$$

Let $\Phi : D \to \mathbb{R}^2$ be a function, holomorphic with respect to the metric ds_Ω and

$$\ln|\Phi(x,y)| \leq -\nu(t) \quad \text{for all} \quad (x,y) \in S_D(0,t) \quad (0 < t < 1) \tag{8.57}$$

for a positive nondecreasing continuous function ν.

Let $\tau = \theta(t)$ be a decreasing function, continuous in $(0,1)$, $\theta(+0) = +\infty$, and

$$\pi \frac{\ln 1/t}{\alpha\,\mu_*(t)} + C_1\,\mu^*(t) \leq \theta(t) \quad (0 < t < 1) \tag{8.58}$$

where

$$C_1 = \pi\alpha\ln 2\left(\frac{2}{\ln^2 2} + \frac{1}{\alpha^2}\right) \qquad \mu^*(t) = \sup_{B_D^{**}(t)} \max\left(\delta, \frac{1}{\delta}\right).$$

If ν satisfies (8.40), then $\Phi(x,y) \equiv 0$.

For the **proof** we use Theorem 8.37. Because $|\nabla h| \equiv 1$, then we are right to put $\nabla h(x,y) = (\cos\beta, \sin\beta)$. By (8.13) we have

$$\lambda_h(t) = \int\limits_{S_D(0,t)} (g^{11}\cos^2\beta + 2g^{12}\cos\beta\sin\beta + g^{22}\sin^2\beta)\sqrt{g}\,\sqrt{dx^2 + dy^2}$$

$$\leq \int\limits_{S_D(0,t)} (\delta + \delta^{-1})\sqrt{dx^2 + dy^2}.$$

Thus, if (8.56) is valid, then (7.18) is also valid.

Now we remark that for $0 < 2t < 1$ from (8.53), it follows

$$\mathrm{mod}_\Omega\,\mathcal{E}(t) \leq \alpha\ln 2\left(\frac{2}{\ln^2 2} + \frac{1}{\alpha^2}\right)\mu^*(t).$$

Moreover, combining estimates (8.44) and (8.54), we find

$$\mathrm{mod}_\Omega\Lambda(t) \leq \frac{\ln 1/t}{\alpha\,\mu_*(t)}.$$

Hence,

$$\mathrm{mod}_\Omega\Lambda(t) + \mathrm{mod}_\Omega\mathcal{E}(t) \leq \frac{\ln 1/t}{\alpha\,\mu_*(t)} + \alpha\ln 2\left(\frac{2}{\ln^2 2} + \frac{1}{\alpha^2}\right)\mu^*(t),$$

and (8.58) implies (8.39). $\qquad\qquad\square$

8.8   Proof of Theorem 8.7

Choose a constant $k > 1$ such that

$$k\frac{\pi}{\alpha} < s.$$

By Lemma 8.15 the complex potential ζ is holomorphic in the domain D. By (8.8) we see that $\varphi \equiv 0$ on γ_0 and the conjugate function ψ defined by (8.11) is an identical constant on γ_α. Denote by C this constant and consider the holomorphic function $\zeta_1 = \zeta - i\,C$.

For every $0 < r < 1$, using (8.9), we have

$$\sup_{S_D(0,r)} |\zeta_1| \quad \leq \operatorname{osc}(\varphi, S_D(0,r)) + \operatorname{osc}(\psi - C, S_D(0,r)) \leq$$

$$\leq \int_{S_D(0,r)} |\nabla\varphi| \sqrt{dx^2 + dy^2} + \int_{S_D(0,r)} |\nabla\psi| \sqrt{dx^2 + dy^2} \leq$$

$$\leq \alpha r \exp\left\{-\frac{1}{r^s}\right\} + \alpha r \exp\left\{-\frac{1}{r^s}\right\} \sup_{S_D(0,r)} \delta(|\nabla\varphi|).$$

The condition (8.9) implies that $\nabla\varphi(x,y) \to 0$ as $(x,y) \to 0$. Because the coefficient $\delta(q)$ in (8.1) is a continuous function and $\delta(0) = 0$, then we can assume that for sufficiently small $t > 0$, quantities

$$\sup_{S_D(0,t)} \delta(|\nabla\varphi|), \quad \frac{1}{\mu_*(t)}, \quad \text{and} \quad \mu^*(t)$$

are not greater than k. Thus for sufficiently small $t > 0$ it is fulfilled

$$\sup_{S_D(0,t)} |\zeta_1| \leq 2\,\alpha\,t\,k \, \exp\left\{-\frac{1}{t^s}\right\} \leq \exp\left\{-\frac{1}{t^{s_1}}\right\}$$

where s_1 is a constant,

$$k\frac{\pi}{\alpha} < s_1 < s.$$

We will use Theorem 8.55. The assumption (8.56) follows from the boundedness of $\delta + \delta^{-1}$ in a neighborhood of 0. The condition (8.57) follows from (8.9) with the function

$$\nu(t) = \frac{1}{t^{s_1}}.$$

Further we find

$$\pi\frac{\ln 1/t}{\alpha\,\mu_*(t)} + C_1\,\mu^*(t) \leq \pi\,k\frac{\ln 1/t}{\alpha} + C_1\,k \equiv \theta(t).$$

Remark now that for the inverse function

$$t = \theta^{-1}(\tau) = \exp\left\{-\frac{\alpha}{\pi}\frac{\tau - C_1 k}{k}\right\},$$

it is valid

$$\int\limits^{+\infty} \nu\left(\theta^{-1}(\tau)\right) e^{-\tau}\, d\tau = \int\limits_0 d\tau \left/\tau^{1+s_1 - \frac{k\pi}{\alpha}}\right. = +\infty.$$

From this we verify (8.40).

This statement follows now from Theorem 8.55.   □

8.9 Half-strips

Consider the half-strip $\Pi = \{(x,y) : 0 < x < \infty,\ 0 < y < \hbar\}$. Fix prime ends $e'' = (0,0)$, $e_0 = (0,1)$, $e' = \infty$ and the function

$$h(x) = \frac{1}{1+x}.$$

This function satisfies (7.16) and (8.20). For an arbitrary $0 < t < +\infty$, we denote

$$S_\Pi(t) = \Pi \cap \{(x,y) : h(x) = t\}, \quad B_\Pi(t) = \Pi \cap \{(x,y) : h(x) < t\}.$$

We estimate $\mathrm{mod}_\Omega \mathcal{E}(t)$. Fix constants $\Delta > 0$ and $t > 0$ such that $0 < \Delta < t$. Denote by $\mathcal{E}_1(t)$ the set of all arcs $\gamma \in \mathcal{E}(t)$ lying in $B_\Pi(t+\Delta) \setminus B_\Pi(t-\Delta)$ and by $\mathcal{E}_2(t)$ - the set of arcs $\gamma \in \mathcal{E}(t) \setminus \mathcal{E}_1(t)$. We have

$$\mathrm{mod}_\Omega \mathcal{E}(t) \leq \mathrm{mod}_\Omega \mathcal{E}_1(t) + \mathrm{mod}_\Omega \mathcal{E}_2(t). \tag{8.59}$$

Setting here $Q_\Pi^*(t) = B_\Pi(t+\Delta) \setminus B_\Pi(t-\Delta)$ and

$$\rho(x,y) = \begin{cases} 1 & \text{for} \quad (x,y) \in Q_\Pi^*(t), \\ 0 & \text{outside of} \quad Q_\Pi^*(t), \end{cases}$$

as above in Section 8.7, we obtain

$$\mathrm{mod}_\Omega \mathcal{E}_1(t) \leq \frac{\displaystyle\iint_{Q_\Pi^*(t)} \max\left(\delta, 1/\delta\right)\, dx\, dy}{\left(\displaystyle\inf_\gamma \int\limits_{\gamma \in \mathcal{E}_1(t)} \sqrt{dx^2 + dy^2}\right)^2}.$$

Since for every arc $\gamma \in \mathcal{E}_1(t)$, it is fulfilled

$$\int_{\gamma \in \mathcal{E}_1(t)} \sqrt{dx^2 + dy^2} \geq \hbar,$$

we arrive to the estimate

$$\text{mod}_\Omega \mathcal{E}_1(t) \leq \frac{1}{\hbar^2} \iint_{Q_\Pi^*(t)} \max(\delta, 1/\delta) \, dxdy. \qquad (8.60)$$

To estimate $\text{mod}_\Omega \mathcal{E}_2(t)$, we observe that every arc $\gamma \in \mathcal{E}_2(t)$ intersects simultaneously $S_\Pi(t)$ and at least one arc $S_\Pi(t - \Delta)$, $S_\Pi(t + \Delta)$. From this we conclude that

$$\text{mod}_\Omega \mathcal{E}_2(t) \leq \text{mod}_\Omega \mathcal{E}_3(t) + \text{mod}_\Omega \mathcal{E}_4(t) \qquad (8.61)$$

where

$$\mathcal{E}_3(t) = \{\gamma \in \mathcal{E}(t) : \gamma \cap S_\Pi(t - \Delta) \neq \emptyset\}$$

and

$$\mathcal{E}_4(t) = \{\gamma \in \mathcal{E}(t) : \gamma \cap S_\Pi(t + \Delta) \neq \emptyset\}.$$

There exists a subarc γ^* on every arc $\gamma \in \mathcal{E}_3(t)$, lying in the domain $\Pi \cap \{t - \Delta < h(x, y) < t\}$ and joining $S_\Pi(t - \Delta)$ with $S_\Pi(t)$. Let $\mathcal{E}_3^*(t)$ be a family of all such arcs γ^*. It is easy to see that

$$\text{mod}_\Omega \mathcal{E}_3(t) \leq \text{mod}_\Omega \mathcal{E}_3^*(t).$$

Choose in (8.46) the density $\rho(x, y) = 1$ for $(x, y) \in Q_\Pi^{**}(t)$ where $Q_\Pi^{**}(t) = B_\Pi(t) \setminus B_\Pi(t - \Delta)$, and define it by zero in all other points. We find

$$\text{mod}_\Omega \mathcal{E}_3^*(t) \leq \frac{\iint_{Q_\Pi^{**}(t)} \max(\delta, 1/\delta) \, dxdy}{\left(\inf_{\gamma} \int_{\gamma^* \in \mathcal{E}_3^*(t)} \sqrt{dx^2 + dy^2} \right)^2} \leq$$

$$\leq \Delta^{-2} \iint_{Q_\Pi^{**}(t)} \max(\delta, 1/\delta) \, dxdy.$$

Thus

$$\text{mod}_\Omega \mathcal{E}_3(t) \leq \Delta^{-2} \iint_{Q_\Pi^{**}(t)} \max(\delta, 1/\delta) \, dxdy \qquad (8.62)$$

Analogously,

$$\text{mod}_\Omega \mathcal{E}_4(t) \leq \Delta^{-2} \iint_{Q_\Pi^{**}(t)} \max(\delta, 1/\delta) \, dxdy \qquad (8.63)$$

Combining now (8.59)–(8.63), we arrive to the following statement:

Lemma 8.64   *If $0 < 2t < 1$, then*

$$\mathrm{mod}_\Omega \mathcal{E}(t) \leq \left(\frac{2}{\Delta^2} + \frac{1}{\hbar^2}\right) \iint_{Q_\Pi^{**}(t)} \max\left(\delta, \frac{1}{\delta}\right) dx dy. \qquad (8.65)$$

Estimate $\mathrm{mod}_\Omega \Lambda(t)$. As above, we will use Lemma 8.43. Let

$$\delta_*(t) = \inf_{\Pi \setminus B_\Pi(t)} \min\left(\delta, 1/\delta\right).$$

We have

$$\inf_\xi \iint_{\Pi \setminus B_\Pi(t)} \min\left(\delta, \frac{1}{\delta}\right) |\nabla \xi|^2 \, dx dy \geq \delta_*(t) \inf_\xi \iint_{\Pi \setminus B_\Pi(t)} |\nabla \xi|^2 \, dx dy =$$

$$= \delta_*(t) \, \mathrm{cap}(S_\Pi(t), \widetilde{e_0 e''}; \Pi \setminus B_\Pi(t)).$$

Since the domain Π is a half-strip of the width $\hbar > 0$ and the domain $\Pi \setminus B_\Pi(t)$ is a rectangle of the length $(1 - t)/t$, then

$$\mathrm{cap}(S_\Pi(t), \widetilde{e_0 e''}; \Pi \setminus B_\Pi(t)) = \frac{\hbar}{(1 - t)/t}.$$

From here we arrive to the estimate:

$$\inf_\xi \iint_{\Pi \setminus B_\Pi(t)} \min\left(\delta, \frac{1}{\delta}\right) |\nabla \xi|^2 \, dx dy \geq \frac{\hbar t \, \delta_*(t)}{1 - t}. \qquad (8.66)$$

Theorem 8.67   *Let Π be a half-strip of the width $0 < \hbar < +\infty$. Let φ be a generalized solution of (8.1) satisfying (8.18) and such that*

$$\int_0^\cdot t^4 \, dt \Big/ \int_{S_\Pi(t)} \frac{\varphi_y^2 + \delta^2 \, \varphi_x^2}{\delta \, |\nabla \varphi|^2} \, |dy| = +\infty. \qquad (8.68)$$

Let $\Phi : \Pi \to \mathbb{R}^2$ be a function holomorphic with respect to a metric ds_Ω; moreover,

$$\ln |\Phi(x, y)| \leq -\nu(t) \quad \text{for all} \quad (x, y) \in S_\Pi(t) \quad (0 < t < 1) \qquad (8.69)$$

for a positive nondecreasing continuous function ν.

Let $\tau = \theta(t)$ be a strictly monotone decreasing, continuous on $(0,1)$ function, $\theta(+0) = +\infty$ and

$$\pi \frac{1-t}{\hbar\, t\, \delta_*(t)} + C_2\, \delta^*(t) \le \theta(t) \quad (0 < t < 1) \tag{8.70}$$

where

$$C_2 = \pi \hbar \Delta \left(\frac{2}{\Delta^2} + \frac{1}{\hbar^2}\right) \qquad \delta^*(t) = \sup_{B_{\Pi}^{**}(t)} \max\left(\delta, \frac{1}{\delta}\right).$$

If ν satisfies (8.40), then $\Phi(x,y) \equiv 0$.

Proof. We use Theorem 8.37. We have

$$\nabla h = (-1/(1+x)^2, 0).$$

By (8.13)

$$\lambda_h(t) = \frac{1}{(1+x)^4} \int\limits_{S_{\Pi}(t)} g^{11}\sqrt{g}\sqrt{dx^2 + dy^2} \le$$

$$\le \frac{1}{(1+x)^4} \int\limits_{S_{\Pi}(t)} \frac{\varphi_y^2 + \delta^2\, \varphi_x^2}{\delta\, |\nabla\varphi|^2} |dy|.$$

Then (7.18) is valid if (8.68) is valid.

From (8.65) it follows that

$$\mathrm{mod}_\Omega\, \mathcal{E}(t) \le \hbar\Delta \left(\frac{2}{\Delta^2} + \frac{1}{\hbar^2}\right) \delta^*(t).$$

Combining estimates (8.44) and (8.66), we obtain

$$\mathrm{mod}_\Omega\Lambda(t) \le \frac{1-t}{\hbar\, t\, \delta_*(t)}.$$

Thus

$$\mathrm{mod}_\Omega\Lambda(t) + \mathrm{mod}_\Omega\mathcal{E}(t) \le \frac{1-t}{\hbar\, t\, \delta_*(t)} + \hbar\Delta \left(\frac{2}{\Delta^2} + \frac{1}{\hbar^2}\right) \delta^*(t),$$

and by (8.70) we conclude that (8.39) is true. $\qquad\qquad\qquad\qquad\square$

8.10   Proof of Theorem 8.4

Choose a constant $k > 1$ such that

$$k\frac{\pi}{\hbar} < s.$$

By Lemma 8.15 the complex potential Ω is holomorphic in Π. By (8.5) we see that $\varphi \equiv 0$ for $y = \hbar$ and $\psi \equiv C$ ($C = \text{const}$) for $y = 0$. We set $\Omega_1 = \Omega - iC$.

For every $0 < t < 1$ by (8.6), we have

$$\sup_{S_\Pi(t)} |\zeta_1| \; \le \text{osc}(\varphi, S_\Pi(t)) + \text{osc}(\psi - C, S_\Pi(t)) \le$$

$$\le \int_{S_\Pi(t)} |\nabla\varphi|\,|dy| + \int_{S_\Pi(t)} |\nabla\psi|\,|dy| \le$$

$$\le \hbar\exp\left\{-\exp\{st\}\right\} + \hbar\exp\left\{-\exp\{st\}\right\}\sup_{S_\Pi(t)}\delta(|\nabla\varphi|).$$

From (8.6) it follows that $\nabla\varphi(x, y) \to 0$ as $x \to +\infty$. Since the coefficient $\delta(q)$ in (8.1) is continuous and $\delta(0) = 0$, then for sufficiently small $t > 0$, quantities

$$\sup_{S_\Pi(t)}\delta(|\nabla\varphi|), \quad \delta^*(t) \quad \text{and} \quad \frac{1}{\delta_*(t)}$$

are not greater than k. Thus for small $t > 0$, we obtain

$$\sup_{S_\Pi(t)} |\zeta_1| \le 2\hbar k \exp\left\{-\exp\{st\}\right\}.$$

We will use Theorem 8.37. The relation (8.68) as $x = +\infty$ is obvious. The assumption (8.69) follows from (8.6) with the function

$$\nu(t) = -\exp\frac{s(1-t)}{t}.$$

Further, we find

$$\pi\frac{1-t}{\hbar t\,\delta_*(t)} + C_2\,\delta^*(t) \le \pi\,k\frac{1-t}{\hbar t} + C_2\,k \equiv \theta(t).$$

Remark now that for the inverse function

$$t = \theta^{-1}(\tau) = \left(1 + \frac{\hbar}{\pi k}(\tau - C_2 k)\right)^{-1},$$

it is fulfilled (8.40).

The necessary statement follows from Theorem 8.37. $\square$

Open questions 8.71 1) Consider solutions stabilization in angle and strip regions of the general form. 2) Consider the problem of the admissible stabilization speed for solutions of nonlinear equations of the high order.

Chapter 9

Critical Points of a Solution

Here we prove an analog of a Nitsche Theorem for solutions of minimal surface type equations in narrow domains and obtain some estimates of the summary topological index of critical points.

9.1 The Nitsche Problem

The following Nitsche Theorem [137] is well-known. *Let $D \subset \mathbb{R}^2$ be a corner of angle $\alpha > 0$. Let $u(x)$ be a solution of minimal surface equation (2.20) in a domain D with zero Dirichlét condition on the boundary ∂D. If $\alpha < \pi$, then $u(x) \equiv 0$ in D.*

Emphasize that for $0 < \alpha < \pi$, we do not assume a priori restrictions to the growth of solutions. Thus, instead of the classical alternative of the Phragmén-Lindelöf type here, we have an essentially more strong statement. For $\alpha = \pi$, the Phragmén-Lindelöf principle is valid in the standard form (see [61], [194], [198]).

The proof of the statement is simple. Nitsche had noticed that for $\alpha < \pi$, there exist some special solutions $\varphi(x)$ of (2.20), vanishing on sides of the corner D and having the infinite gradient on some arcs, joining sides of D. These considerations are completed with the application of the maximum principle for the difference of solutions $u(x) - \varphi(x)$.

Nitsche [137] constructed the special solutions $\varphi(x)$ for corners D with the Enneper surface. In the general case, the problem of the existence such solutions is very difficult. This is the most principal obstacle — to prove the showed statement for solutions in domains with curvilinear boundaries of the general form.

195

In [113], we had suggested another explanation of the Nitsche effect, suitable for domains with curvilinear boundaries. Briefly, our approach consists of the following. Let $D \subset \mathbb{R}^2$ be a simply connected domain. We put

$$v(x) = \int_{x_0}^{x} -\frac{u'_{x_2}}{\sqrt{1 + |\nabla u|^2}} dx_1 + \frac{u'_{x_1}}{\sqrt{1 + |\nabla u|^2}} dx_2, \quad x_0, x \in D.$$

This function is univalent, and $u(x) + i\,v(x)$ is holomorphic with respect to the metric of the surface, described by the equation $x_3 = u(x_1, x_2)$. Suppose that the domain D is unbounded and $u(x) = 0$ on ∂D. Suppose also that we can prove the Phragmén-Lindelöf principle for $u(x) + i\,v(x)$. Then the function $v(x)$ must be either an identical constant or to have a sufficiently rapid growth as $x \to \infty$ along the domain D. Moreover, the growth of $v(x)$ can be arbitrarily rapid if the domain D is sufficiently narrow in the neighborhood of the infinite point of $\mathbb{R}^2$. However, $|\nabla v(x)| < 1$ in D, and $v(x)$ can not growth too rapid (for example, if D is a corner then $v(x)$ can not growth faster than linear functions). Thus, *the Nitsche statement should be valid in domains in which the hypothetical Phragmén-Lindelöf principle for $v(x)$ in D contradicts the restriction $|\nabla v(x)| < 1$ in D.*

The described scheme is effective as in the case of zero Dirichlét boundary conditions as in the case of zero Neyman conditions and extends to solutions of the minimal surface type equation.

An analog of Nitsche theorem for solutions of minimal surface type equations permits to prove the following statement.

Every entire[1] solution $u(x)$ of a minimal surface type equation has finite number of critical points $a_1, a_2, \ldots, a_N$. Moreover,

$$\sum_{i=1}^{N} i(a_i) \leq c(\nu)$$

where $i(a_i)$ is the topological index of the function $u(x)$ at a_i, and $c(\nu)$ is a quantity, defined only the equation.

9.2 Generalized Solutions

At first, we describe the class of equations. Let $D \subset \mathbb{R}^2$ be a domain and let $A : D \times \mathbb{R}^2 \to \mathbb{R}^2$ be a map satisfying the following assumptions:

(*i*) for a.e. $x \in D$, the map $\xi \in \mathbb{R}^2 \to A(x, \xi)$ is defined and continuous;

---

[1]i.e. defined in the plane $\mathbb{R}^2$.

(*ii*) the map $x \in D \to A(x, \xi)$ is Lebesgue measurable for all $\xi \in \mathbb{R}^2$;

(*iii*) for a.e. $x \in D$ and all $\xi \in \mathbb{R}^2$, the following structure restrictions hold

$$\nu_1 \frac{|\xi|^2}{\sqrt{1 + |\xi|^2}} \le \langle \xi, A(x, \xi) \rangle, \tag{9.1}$$

$$|A(x, \xi)| \le \nu_2 \frac{|\xi|}{\sqrt{1 + |\xi|^2}}. \tag{9.2}$$

Here, ν_1, $\nu_2 > 0$ are constants.

We denote $\nu = \nu_2 / \nu_1$. It is clear that always $\nu \ge 1$.

The assumptions (i) and (ii) guarantee the measurability of $x \in D \to A(x, u(x))$ for every measurable on D vector field $u(x)$ (see details in [59, Section **3**]).

Consider the equation

$$L[u] \equiv \operatorname{div} A(x, \nabla u) = 0. \tag{9.3}$$

We need to define generalized solutions of the equation (9.3) with zero mixed boundary conditions. Technically it is convenient to use the following definition.

Let $D \subset \mathbb{R}^2$ be a domain and let $E \subset \partial \tilde{D}$ be a closed set of prime ends of D. Let $\Delta \subset \tilde{D}$ be a set and $[\Delta]_{\tilde{D}}$ its closuring with respect to topology of $\tilde{D}$. We say that $[\Delta]_{\tilde{D}} \subset (D \cup E)$ if every infinite sequence of points $x_n \in \Delta$, not having limiting points in D, does not have limiting points (with respect to the topology $\tilde{D}$) on $\partial \tilde{D} \setminus E$. Further, for an arbitrary function $h : D \to \mathbb{R}$, we put

$$\operatorname{supp}_{\tilde{D}} h \equiv [\{x : h(x) \ne 0\}]_{\tilde{D}}.$$

A function $u \in \operatorname{Lip}_{\mathrm{loc}} D$ is called the *generalized solution* of the equation (9.3) with zero mixed boundary conditions on the set $E \subset \partial \tilde{D}$ if, for every function $\varphi \in \operatorname{Lip}_{\mathrm{loc}} D$ such that

$$\operatorname{supp}_{\tilde{D}} \varphi \subset (D \cup E), \tag{9.4}$$

it is fulfilled

$$\iint\limits_{D} \langle \nabla \varphi, A(x, \nabla u) \rangle \, dx_1 \, dx_2 = 0. \tag{9.5}$$

Consider the following example. Let D be a simply connected unbounded domain in $\mathbb{R}^2$ with the smooth boundary ∂D. Suppose that $u \in C^2(D) \cap C^1(\overline{D})$, the vector function $A(x, \xi)$ belongs to the class $C^1(D \times \mathbb{R}^2)$ and that at every boundary point $x \in \partial D$, it is fulfilled at least at one relation

$$u(x) = 0, \quad \langle A(x, \nabla u), \mathbf{n} \rangle = 0 \qquad (9.6)$$

where $\mathbf{n}$ is an unit vector of the inner normal to ∂D.

In this case the space $\tilde{D}$ coincides with the closure of D as a domain on the Riemannian sphere $\tilde{\mathbb{R}}^2$, and we can put

$$E = \partial D = \tilde{\partial} D \setminus \{\infty\}.$$

Let $\varphi = u\psi$ where the function $\psi \in C_0^1(\mathbb{R}^2)$ is arbitrary. The function φ has the support $\operatorname{supp}_{\tilde{D}} \varphi \subset (D \cup \partial D)$, and thus, the relation (9.4) is valid for φ. By the Green formula, we have

$$\int\limits_{\partial D} \psi(x) u(x) \, \langle A(x, \nabla u), \mathbf{n} \rangle \, |dx| \;=\; \iint\limits_{D} \langle \nabla(\psi u), A(x, \nabla u) \rangle \, dx_1 \, dx_2 +$$

$$+ \;\iint\limits_{D} \psi(x) u(x) \, L[u] \, dx_1 \, dx_2 \,.$$

From this, by conditions (9.6) and the equality $L[u] = 0$, we arrive to (9.5).

Conversely, under assumptions $L[u] = 0$ and (9.5), the realization of the relation

$$\int\limits_{\partial D} \psi(x) \, u(x) \, \langle A(x, \nabla u), \mathbf{n} \rangle \, |dx| = 0$$

with the described arbitrariness in the choice of $\psi(x)$ implies (9.6).

It is clear that the minimal surface equation (2.20) is contained in the described class of equations. However, it can not possess some properties, specific for solutions of the equation (2.20). For example, famous Bernstein theorem states that *every entire solution of the minimal surface equation is a linear function* [19, p. 257]. It is not difficult to check that the equation

$$\sum_{i=1}^{2} \frac{\partial}{\partial x_i} \left(\frac{u_{x_i}}{\sqrt{1 + u_{x_i}^2}} \right) = 0$$

satisfies all restrictions (9.1), (9.2). Nevertheless, it has entire solutions $u = C\, x_1 x_2$ which are not linear.

9.3   Maps onto a Half-plane

Let D be a simply connected unbounded domain in $\mathbb{R}^2$ with the boundary ∂D, rectifiable for every compact portion. Below we will call such domains *elementary*.

Let $u(x)$ be a generalized solution of an equation $L[u] = 0$ in D satisfying the condition

$$u \in \mathrm{Lip}\,(D_r), \quad D_r = \{x \in D : |x| < r\} \tag{9.7}$$

for every $r > 0$.

Let Ω be the graph of the solution $x_3 = u(x_1, x_2)$. As above, we denote by

$$ds_\Omega^2 = \sum_{i,j=1}^{2} (\delta_{ij} + u_{x_i} u_{x_j})\, dx_i dx_j$$

the length element on Ω, and by

$$d\Omega = \sqrt{1 + |\nabla u|^2}\, dx_1 dx_2$$

- the area element of Ω.

As above in Section 3.7, we introduce isothermal coordinates on the surface Ω. Namely, let $x \in D$ be a point in which $u(x)$ is differentiable. Since the quadratic form ds_Ω^2 is positively defined in D, then the infinitesimal disc with respect to the metric ds_Ω^2 with the center at x is an infinitesimal ellipse with respect to the Euclidean metric. Let $\theta(x)$, $p(x)$ be characteristics of this ellipse. Because Ω is the graph of the locally Lipschitz function, then $p(x)$ is locally bounded in D. By Theorem 3.49 there exists a quasi-conformal map $\xi = \xi(x)$ of D into the $\xi = (\xi_1, \xi_2)$-plane, characteristics of which coincide with $\theta(x), p(x)$ a.e. in D. This map is defined up to a conformal transformation in the ξ-plane.

Denote by $x = x(\xi) = (x_1(\xi), x_2(\xi))$ the map, inverse to $\xi = \xi(x)$, and by x_3 — the function $u(x(\xi))$. The vector function

$$\chi(\xi) = (x_1(\xi), x_2(\xi), x_3(\xi))$$

realizes the schlicht conformal map of the domain $D^* = \xi(D)$ onto the surface Ω and defines isothermal coordinates on Ω.

Because the solution $u(x)$ satisfies (9.7), then the characteristic $p(x)$ is bounded in a neighborhood of every finite boundary point of the domain D and can indefinitely grow only as $x \to \infty$ along D. Using Corollary 4.26 it is not difficult to conclude that D^* is a simply connected domain different from $\mathbb{R}^2$.

By an auxiliary conformal map in the $\xi = (\xi_1, \xi_2)$-plane, we attain that the image D^* of D under the map $\xi = \xi(x)$ is the upper half-plane $\{\xi \in \mathbb{R}^2 : \xi_2 >$

0}. The boundary ∂D is a simple Jordan arc and the assumption (9.7) holds; therefore, by Theorem 4.24 the map $\xi = \xi(x) : D \to D^*$ is continuous up to ∂D. Moreover, the extended map is one-to-one on this arc.

However, *it is not clear a priory what is the image of the infinite point on the boundary of D: is it a unit prime end of D^* or a continuum of prime ends?*

Fix the points $a, b \in \partial D$ and suppose that the map $\xi = \xi(x)$ satisfies assumptions

$$\xi(a) = (0,0), \quad \xi(b) = (1,0)$$

and that there exists a sequence of points $x_k \in D$, $x_k \to \infty$, along which $\xi(x_k) \to \infty$ $(k \to \infty)$.

By foregoing constructions such normalization is possible.

We put

$$m(t) = \min_{|x-a|=t} |\xi(x)|.$$

For an arbitrary $t > 0$, we denote by $S_D(t)$ a component of

$$\{x \in D : |x - a| = t\}$$

separating boundary points a and ∞ in D. By the symbol $D(r, R)$, we denote the part of D lying between $S_D(r)$ and $S_D(R)$ for $0 < r < R < \infty$.

The following statement permits to conclude that the image of D under the map $\xi = \xi(x)$ is whole upper half-plane.

Theorem 9.8 *Let $u(x)$ be a generalized solution of (9.3) on an elementary, unbounded, simply connected domain D. Suppose that $u(x)$ satisfies a zero mixed boundary condition on ∂D. Then the map $\xi = \xi(x)$ satisfies the condition*

$$\lim_{x \to \infty} \xi(x) = \infty.$$

Moreover,

$$\lim_{t \to \infty} \frac{\ln m(t)}{\ln^2 t} \left(\iint_{D(1,t)} \frac{dx_1 dx_2}{|x|^2} + 2\nu\pi^2 \right) \geq \frac{2}{\nu}. \tag{9.9}$$

9.3.1 Estimates of a Module

Without loss of generality, we can assume that a point a is the origin in $\mathbb{R}^2$. Denote by $\Gamma_\Omega(r, R)$ the family of arcs γ, lying in $D(r, R)$ and joining $S_D(r)$ and $S_D(R)$. The most principal place in constructions supplies an estimate of the module of the family $\Gamma(r, R)$ with respect to the metric ds_Ω^2. The specific of this estimate consists in the uncertainty of metric coefficients a priori. In other words, we estimate module of condensers on the unknown surface. There is a significant difference between the considered situation and the standard case in which we estimate the module of such condenser in $\mathbb{R}^2$.

Lemma 9.10   *The following inequality holds*

$$\mathrm{mod}_\Omega \Gamma(r, R) \leq \frac{\nu\pi}{2\ln^2\left(1 + \dfrac{R-r}{M}\right)} \left(\iint_{D(r,R)} \frac{dx_1 dx_2}{|x|^2} + 2\nu\pi^2\right) \qquad (9.11)$$

where

$$M = \max_{S(r)} \left(|x|^2 + u^2(x)\right)^{-1/2}.$$

Proof. We will use Theorem 1.14. In (1.15) choose the density

$$\rho(x) = \left(|x|^2 + u^2(x)\right)^{-1/2}$$

for $x \in D(r, R) \cap \{r < |x| < R\}$ and $\rho(x) = 0$ for all remaining $x \in D$. Then we have

$$\mathrm{mod}_\Omega \Gamma(r, R) \leq \frac{\displaystyle\iint_{D(r,R)} \left(|x|^2 + u^2(x)\right)^{-1} d\Omega}{\left(\displaystyle\inf_{\gamma \in \Gamma(r,R)} \int_\gamma \left(|x|^2 + u^2(x)\right)^{-1/2} ds_\Omega\right)^2}. \qquad (9.12)$$

Observing that on every arc $\gamma \in \Gamma(r, R)$ there exists a subarc γ', lying in $D(r, R) \cap \{r < |x| < R\}$ and having endpoints on boundary arcs $S_D(r)$ $S_D(R)$, we estimate the denominator

$$\int_\gamma \left(|x|^2 + u^2(x)\right)^{-1/2} ds_\Omega \geq \int_{\gamma'} \left(|x|^2 + u^2(x)\right)^{-1/2} \left|d\left(|x|^2 + u^2(x)\right)^{-1/2}\right|,$$

and further,

$$\int_\gamma \left(|x|^2 + u^2(x)\right)^{-1/2} ds_\Omega \geq \ln\left(1 + \frac{R-r}{M}\right). \qquad (9.13)$$

Now we estimate the numerator in the right side of the inequality (9.12). We put

$$\eta(x) = \frac{1}{|x|}\mathrm{arctg}\,\delta(x), \quad \delta(x) = \frac{u(x)}{|x|}.$$

Fix arbitrarily $R' > R$, $0 < r' < r$ and denote by $\psi(\tau)$ the function

$$
\psi(\tau) = \begin{cases}
0 & \text{for} \quad \tau \in [0, r') \,; \\[2mm]
(\tau - r')/(r - r') & \text{for} \quad \tau \in [r', r) \,; \\[2mm]
1 & \text{for} \quad \tau \in [r, R) \,; \\[2mm]
(R' - \tau)/(R' - R) & \text{for} \quad \tau \in [R, R') \,; \\[2mm]
0 & \text{for} \quad \tau \in [R', \infty] \,.
\end{cases}
$$

The function $\varphi(x) = \psi(|x|)\,\eta(x)$ belongs to the class $\mathrm{Lip}_{\mathrm{loc}}\, D$ and has the property

$$
\mathrm{supp}_{\tilde{D}}\,\varphi \subset \{x : r' < |x| \le R'\} \cap \mathrm{supp}_{\tilde{D}}\, u \subset (D \cup \partial D) \,.
$$

Therefore, it is admissible in the integral relation (9.5), and we can write

$$
\iint\limits_D \eta(x)\, \langle \nabla\psi, A(x, \nabla u)\rangle\, dx_1\, dx_2 = -\iint\limits_D \psi(x)\, \langle \nabla\eta, A(x, \nabla u)\rangle\, dx_1\, dx_2 \,.
$$

Thus we find

$$
\iint\limits_D \psi(x)\, \langle \nabla\eta, A(x, \nabla u)\rangle\, dx_1\, dx_2 \le \iint\limits_D |\eta(x)|\, |\nabla\psi|\, |A(x, \nabla u)|\, dx_1\, dx_2 \,.
$$

We remark that

$$
\nabla\eta = \frac{\nabla u}{|x|^2 + u^2(x)} - \frac{\nabla|x|}{|x|^2}\left(\mathrm{arctg}\,\delta + \frac{\delta}{1 + \delta^2}\right),
$$

and also

$$
\langle \nabla u, A(x, \nabla u)\rangle \ge 0 \,, \quad |\eta(x)| \le \frac{\pi}{2\,|x|} \,,
$$

and

$$
|\nabla\psi| = \begin{cases}
1/(R' - R) & \text{for} \quad R < |x| < R' \,, \\[2mm]
1/(r - r') & \text{for} \quad r' < |x| < r \,.
\end{cases}
$$

Thus denoting by

$$
\mathcal{U} = \{x \in D : r' < |x| < r\} \,,
$$

$$\mathcal{V} = \{x \in D : r < |x| < R\},$$
$$\mathcal{W} = \{x \in D : R < |x| < R'\},$$

we arrive to the relation

$$\iint\limits_{\mathcal{V}} \frac{\langle \nabla u, A(x, \nabla u)\rangle dx_1 dx_2}{|x|^2 + u^2(x)} \leq \iint\limits_{\mathcal{V}} \left| \text{arctg}\delta + \frac{\delta}{1+\delta^2} \right| |A(x, \nabla u)| \frac{dx_1\, dx_2}{|x|^2} +$$

$$+ \frac{\pi}{2(r-r')} \iint\limits_{\mathcal{U}} |A(x, \nabla u)| \frac{dx_1\, dx_2}{|x|} + \frac{\pi}{2(R'-R)} \iint\limits_{\mathcal{W}} |A(x, \nabla u)| \frac{dx_1\, dx_2}{|x|}.$$

Structure assumptions (9.1), (9.2) on the differential operator $L[u]$ imply

$$\nu_1 \sqrt{1 + |\nabla u|^2} - \frac{\nu_1}{\sqrt{1 + |\nabla u|^2}} \leq \langle \nabla u, A(x, \nabla u)\rangle,$$

and

$$|A(x, \nabla u)| \leq \frac{\nu_2 |\nabla u|}{\sqrt{1 + |\nabla u|^2}} \leq \nu_2.$$

Thus

$$\iint\limits_{\mathcal{U}} \frac{\sqrt{1 + |\nabla u|^2}}{|x|^2 + u^2(x)} \, dx_1 dx_2 \leq \iint\limits_{\mathcal{V}} Q(x) \frac{dx_1\, dx_2}{|x|^2} +$$

$$+ \frac{\nu\pi}{2(r-r')} \iint\limits_{\mathcal{U}} \frac{dx_1\, dx_2}{|x|} + \frac{\nu\pi}{2(R'-R)} \iint\limits_{\mathcal{W}} \frac{dx_1\, dx_2}{|x|} \qquad (9.14)$$

where

$$Q(x) = \frac{1}{(1+\delta^2)\sqrt{1+|\nabla u|^2}} + \frac{1}{\nu_1} \left| \text{arctg } \delta + \frac{\delta}{1+\delta^2} \right| |A(x, \nabla u)|.$$

It is not difficult to see that

$$\frac{\nu\pi}{2(r-r')} \iint\limits_{\mathcal{U}} \frac{dx_1\, dx_2}{|x|} + \frac{\nu\pi}{2(R'-R)} \iint\limits_{\mathcal{W}} \frac{dx_1\, dx_2}{|x|} \leq 2\nu\,\pi^2.$$

Now we have

$$Q(x) \leq \frac{1}{(1+\delta^2)\sqrt{1+|\nabla u|^2}} + \nu \left| \text{arctg } \delta + \frac{\delta}{1+\delta^2} \right| \frac{|\nabla u|}{\sqrt{1+|\nabla u|^2}}$$

$$\leq \left(\frac{1}{(1+\delta^2)} + \nu^2 \left(\frac{\delta}{1+\delta^2} + \text{arctg } \delta \right)^2 \right)^{1/2} \leq \frac{\nu\pi}{2}.$$

Combining these estimates by (9.14), we obtain

$$\iint\limits_{\mathcal{V}} \frac{d\Omega}{|x|^2 + u^2(x)} \leq \frac{\nu\pi}{2} \iint\limits_{\mathcal{V}} \frac{dx_1 dx_2}{|x|^2} + 2\nu\pi^2 \,.$$

By (9.12), (9.13), we arrive to the inequality (9.11). $\qquad\square$

9.3.2 Proof of Theorem 9.8

Let G' be a subdomain of a domain D separated by an arc $S_D(R)$ from the infinite point of $\mathbb{R}^2$. The subdomain G' abuts to a boundary point $a = (0,0)$, and we can use the results obtained in the Problem A considered in Section 7.3.

By Theorem 7.43 we have

$$\inf_{x \in S_D(R)} |\xi| \geq \exp\{-\pi\, k_\Omega(G')\}\,.$$

However, by Lemma 7.61 it is fulfilled

$$k_\Omega(G') \leq \mathrm{mod}_\Omega\mathcal{E}$$

where $\mathcal{E}$ is the family of all locally rectifiable arcs γ separating the boundary point a from $S_D(R)$ and such that $\gamma \cap S_D(r) \neq \emptyset$. From here we find

$$\inf_{x \in S_D(R)} |\xi| \geq \exp\{-\pi\, k_\Omega(G')\}\,;$$

and since the quantity $C = \mathrm{mod}_\Omega\mathcal{E}$ does not depend on R, then

$$\inf_{x \in S_D(R)} |\xi| \geq C_1\exp\{-\pi\, \mathrm{mod}_\Omega\Gamma(r,R)\}$$

where

$$C_1 = \exp\{-\pi\, C\}\,.$$

We use Lemma 9.10. The estimate (9.11) implies that

$$\inf_{x \in S_D(R)} |\xi| \geq \qquad\qquad\qquad\qquad\qquad\qquad\qquad (9.15)$$

$$\geq C_1\exp\left\{ -\frac{\nu\pi^2}{2\ln^2\left(1 + \frac{R-r}{M}\right)} \left(\iint\limits_{D(r,R)} \frac{dx_1 dx_2}{|x|^2} + 2\nu\pi^2 \right) \right\}\,.$$

The first statement of Theorem 9.8 follows now from the inequality

$$\iint\limits_{D(r,R)} \frac{dx_1 dx_2}{|x|^2} \leq 2\pi\, \ln\frac{R}{r}\,.$$

The relation (9.9) is obtained from (9.15) with the limiting passage. $\qquad\square$

9.4   Estimates of the Speed of Growth Solutions

Let
$$D^* = \{\xi = (\xi_1, \xi_2) \in \mathbb{R}^2 : \xi_2 > 0\}$$
be an upper half-plane, let
$$D^*(t) = \{\xi = (\xi_1, \xi_2) \in \mathbb{R}^2 : \xi_2 > 0, |\xi| < t\}$$
and let
$$w(\xi) = (u(\xi_1, \xi_2), v(\xi_1, \xi_2)) : D^* \to \mathbb{R}^2$$
be a holomorphic function.

Suppose that for every $t > 0$ the second derivative $w''(\xi)$ is bounded in $D^*(t)$. From here, in particular, it follows that the first derivative $w'(\xi)$ belongs to $\mathrm{Lip}\,(\overline{D^*(t)})$ in every half-disc $D^*(t)$ and that the functions $w(\xi)$, $w'(\xi)$ are extendable continuously up to the boundary ∂D^*.

9.4.1   An Inequality for the Energy Integral

We will prove an inequality for the energy integral which is called sometimes the Saint-Venant principle (see, for example, [139], [140]). On other modifications and generalizations for the case of solutions of partial elliptic equations of the general form and also for parabolic and hyperbolic equations, see [190], [191], [138], [112], [168], [182], [68], [99], [63], etc.

Theorem 9.16   *Let H_1 be a closed set on the real axis $\xi_2 = 0$ and $H_2 = \partial D^* \setminus H_1$ - its addition. Let $\chi(t)$ be a function which is equal 1 if $H_1 \cap [-t, t] = \emptyset$ or $H_2 \cap [-t, t] = \emptyset$, and equal $1/2$ if $H_1 \cap [-t, t] \neq \emptyset$, $H_2 \cap [-t, t] \neq \emptyset$.*

Suppose that at every point $\xi = (\xi_1, 0)$ of the boundary ∂D^*, it fulfills at least one of conditions:

$$\begin{cases} u(\xi_1, 0) &= 0\,, \quad for \quad (\xi_1, 0) \in H_1\,; \\[2mm] dv(\xi_1, 0) &= 0\,, \quad for \quad (\xi_1, 0) \in H_2\,. \end{cases} \tag{9.17}$$

Then for all $0 < r < R < \infty$, the following inequality holds

$$I(r) \leq I(R) \exp\left\{ -\frac{1}{2} \int_r^R \chi(t)\, \frac{dt}{t} \right\} \tag{9.18}$$

where

$$I(t) = \iint\limits_{D^*(t)} |\nabla u|^2\, d\xi_1 d\xi_2\,.$$

Proof. We will use an idea ascending to Carleman [26]. Fix arbitrarily $r < t < R$. The vector field $u(\xi)\,\nabla v(\xi)$ satisfies the Lipschitz condition in $\overline{D^*(t)}$ and by Green's formula it is fulfilled

$$\int_{\partial D^*(t)} u\,dv = \iint_{D^*(t)} (u_{\xi_1} v_{\xi_2} - u_{\xi_2} v_{\xi_1})\,d\xi_1 d\xi_2\,. \tag{9.19}$$

This statement is a special corollary of [37, Theorem **4.5.6**].

Using Cauchy-Riemann conditions for the vector function $w = (u,v)$ and boundary conditions (9.17), by (9.19) we deduce

$$\int_{S_{D^*}(t)} u\,dv = \iint_{D^*(t)} |\nabla u(\xi)|^2\,d\xi_1 d\xi_2\,. \tag{9.20}$$

Case A). First we suppose that at least one of the ends of $S_{D^*}(t)$ lies on H_1. We have

$$\left| \int_{S_{D^*}(t)} u(\xi)\,dv \right| \leq \int_{S_{D^*}(t)} |u(\xi)|\,|\langle \nabla u(\xi), \mathbf{n}(\xi)\rangle|\,|d\xi| \leq$$

$$\leq \left(\int_{S_{D^*}(t)} |u(\xi)|^2\,|d\xi| \right)^{1/2} \left(\int_{S_{D^*}(t)} |\langle \nabla u(\xi), \mathbf{n}(\xi)\rangle|^2\,|d\xi| \right)^{1/2}\,.$$

We put

$$\lambda_1\left(S_{D^*}(t)\right) = \frac{\left(\displaystyle\int_{S_{D^*}(t)} |\langle \nabla u(\xi), \tau(\xi)\rangle|^2\,|d\xi| \right)^{1/2}}{\left(\displaystyle\int_{S_{D^*}(t)} |u(\xi)|^2\,|d\xi| \right)^{1/2}}$$

where $\tau(\xi)$ is a unit vector tangent to $S_{D^*}(t)$ at the point $\xi \in S_{D^*}(t)$.

Then we have

$$\left| \int\limits_{S_{D^*}(t)} u(\xi)\, dv \right| \le \frac{1}{\lambda_1\left(S_{D^*}(t)\right)} \left(\int\limits_{S_{D^*}(t)} |\langle \nabla u(\xi), \tau(\xi) \rangle|^2\, |d\xi| \right)^{1/2} \times$$

$$\times \left(\int\limits_{S_{D^*}(t)} |\langle \nabla u(\xi), \mathbf{n}(\xi) \rangle|^2\, |d\xi| \right)^{1/2}.$$

Using the Cauchy inequality

$$2\,a\,b \le a^2 + b^2\,,$$

we find

$$\left| \int\limits_{S_{D^*}(t)} u(\xi)\, dv \right| \le \frac{1}{2\lambda_1\left(S_{D^*}(t)\right)} \times$$

$$\times \left(\int\limits_{S_{D^*}(t)} \left(|\langle \nabla u(\xi), \tau(\xi) \rangle|^2 + |\langle \nabla u(\xi), \mathbf{n}(\xi) \rangle|^2 \right)\, |d\xi| \right).$$

However,

$$|\langle \nabla u(\xi), \tau(\xi) \rangle|^2 + |\langle \nabla u(\xi), \mathbf{n}(\xi) \rangle|^2 = |\nabla u(\xi)|^2\,,$$

and hence,

$$\left| \int\limits_{S_{D^*}(t)} u(\xi)\, dv \right| \le \frac{1}{2\lambda_1\left(S_{D^*}(t)\right)} \int\limits_{S_{D^*}(t)} |\nabla u(\xi)|^2\, |d\xi|. \qquad (9.21)$$

Relations (9.20), (9.21) imply

$$\iint\limits_{D^*(t)} |\nabla u(\xi)|^2\, d\xi_1 d\xi_2 \le \frac{1}{2\lambda_1\left(S_{D^*}(t)\right)} \int\limits_{S_{D^*}(t)} |\nabla u(\xi)|^2\, |d\xi|.$$

It is easy to see that

$$I(t) = \iint\limits_{\{\xi \in D^*:\, |\xi| < t\}} |\nabla u|^2\, d\xi_1 d\xi_2 = \int\limits_0^t dt \int\limits_{S_{D^*}(t)} |\nabla u(\xi)|^2\, |d\xi|,$$

and for a.e. $t \geq 0$ we have

$$\frac{d}{dt} I(t) = \int_{S_{D^*}(t)} |\nabla u(\xi)|^2 \, |d\xi| \,.$$

Thus from the previous inequality, we find

$$I(t) \leq \frac{1}{2\lambda_1 (S_{D^*}(t))} \frac{d}{dt} I(t) \,.$$

Integrating this differential inequality, we obtain

$$\exp \left\{ \int_r^R 2\lambda_1 (S_{D^*}(t)) \, dt \right\} \leq \frac{I(R)}{I(r)} \,. \tag{9.22}$$

Case B). Suppose that both ends of the half-circle $S_{D^*}(t)$ lie on H_2. Then for an arbitrary constant C by (9.17), we have

$$(u - C) \, dv|_{\partial D^*} = 0 \,;$$

and using (9.20), we can write

$$\int_{S_{D^*}(t)} (u - C) \, dv = \iint_{D^*(t)} |\nabla u(\xi)|^2 \, d\xi_1 d\xi_2 \,. \tag{9.23}$$

From here

$$\left| \int_{S_{D^*}(t)} (u(\xi) - C) \, dv \right| \leq \int_{S_{D^*}(t)} |u(\xi)| \, |\langle \nabla u(\xi), \mathbf{n}(\xi) \rangle| \, |d\xi| \leq$$

$$\leq \left(\int_{S_{D^*}(t)} |u(\xi)|^2 \, |d\xi| \right)^{1/2} \left(\int_{S_{D^*}(t)} |\langle \nabla u(\xi), \mathbf{n}(\xi) \rangle|^2 \, |d\xi| \right)^{1/2} \,.$$

We put here

$$\lambda_2 (S_{D^*}(t)) = \frac{\left(\displaystyle\int_{S_{D^*}(t)} |\langle \nabla u(\xi), \tau(\xi) \rangle|^2 \, |d\xi| \right)^{1/2}}{\left(\displaystyle\int_{S_{D^*}(t)} |u(\xi) - C|^2 \, |d\xi| \right)^{1/2}} \,.$$

Further, as above, we deduce

$$\left| \int\limits_{S_{D^*}(t)} (u(\xi) -) \, dv \right| \leq \frac{1}{2\lambda_2 \left(S_{D^*}(t) \right)} \int\limits_{S_{D^*}(t)} |\nabla u(\xi)|^2 \, |d\xi| . \qquad (9.24)$$

Combining relations (9.23), (9.24), we arrive at first to the differential inequality

$$I(t) \leq \frac{1}{2\lambda_2 \left(S_{D^*}(t) \right)} \frac{d}{dt} I(t)$$

and, further, to the relation

$$\exp \left\{ \int\limits_{r}^{R} 2\lambda_2 \left(S_{D^*}(t) \right) \, dt \right\} \leq \frac{I(R)}{I(r)} . \qquad (9.25)$$

We estimate quantities $\lambda_1 \left(S_{D^*}(t) \right)$ and $\lambda_2 \left(S_{D^*}(t) \right)$. We recall the well-known Wirtinger inequality (see, for example, [12, Chapter 5, §**10**] or, in the general case, [11, Chapter III])

$$\int\limits_{0}^{2\pi} h^2(s) \, ds \leq \int\limits_{0}^{2\pi} h'^2(s) \, ds \qquad (9.26)$$

which is valid for every 2π-periodic function $h : \mathbb{R} \to \mathbb{R}$ satisfying the Lipschitz condition and such that

$$\int\limits_{0}^{2\pi} h(s) \, ds = 0. \qquad (9.27)$$

Moreover, the equality in (9.26) is supplied with the function

$$h(\tau) = c_1 \cos s + c_2 \sin s .$$

If the function $h : \mathbb{R} \to \mathbb{R}$ is periodic with a period $T \geq 0$ and the integral along the period equals zero, then obviously,

$$\int\limits_{0}^{T} h^2(s) \, ds \leq \left(\frac{2\pi}{T} \right)^2 \int\limits_{0}^{T} h'^2(s) \, ds .$$

Under the assumptions (A), setting

$$h(s) = u(t \cos s, t \sin s) ,$$

we have

$$|d\xi|_{\xi \in S_{D^*}(t)} = t\, ds\,, \quad |\langle \nabla u(\xi), \tau(\xi)\rangle|^2 = \frac{1}{t^2} h'^2(s)\,,$$

and further,

$$\lambda_1\left(S_{D^*}(t)\right) = \frac{1}{t} \frac{\left(\displaystyle\int_0^{\pi} |h'(s)|^2\, ds\right)^{1/2}}{\left(\displaystyle\int_0^{\pi} h^2(s)\, ds\right)^{1/2}}\,.$$

Two cases are possible. Suppose that both ends of the half-circle $S_{D^*}(t)$ lie on H_1. Then

$$h(0) = h(\pi) = 0\,.$$

We extend the function $h(s)$ at first by the symmetry with respect to the point $s = 0$ up to the function $h^*(s)$, given on $[-\pi, \pi]$ by the rule $h^*(s) = -h^*(-s)$ ($-\pi \le s \le \pi$) and satisfying the condition

$$\int_{-\pi}^{\pi} h^*(s)\, ds = 0\,.$$

Further, we extend it by the periodicity up to the Lipschitz function

$$h^{**} : \mathbb{R} \to \mathbb{R}\,.$$

This function is 2π-periodic, satisfies (9.27), and by the inequality (9.26) we obtain

$$\lambda_1\left(S_{D^*}(t)\right) = \frac{1}{t} \frac{\left(\displaystyle\int_0^{2\pi} |(h^{**})'(s)|^2\, ds\right)^{1/2}}{\left(\displaystyle\int_0^{2\pi} (h^{**})^2(s)\, ds\right)^{1/2}} \ge \frac{1}{t}\,. \tag{9.28}$$

If only one of the ends of the half-circle $S_{D^*}(t)$ lies on H_1, then without loss of generality, we put $h(\pi) = 0$. Next we extend this function by the rule

$$h^*(s) = h^*(-s)\,, \quad h^* : [-\pi, \pi] \to \mathbb{R}\,.$$

Then we have $h^*(-\pi) = h^*(\pi) = 0$, and $h^*(s)$ is extended up to the 4π-periodic function $h^{**}(s)$, satisfying the Lipschitz condition and having the zero integral along the period. Thus we obtain

$$\lambda_1\left(S_{D^*}(t)\right) = \frac{1}{t}\frac{\left(\displaystyle\int\limits_0^{4\pi}|(h^{**})'(s)|^2\,ds\right)^{1/2}}{\left(\displaystyle\int\limits_0^{4\pi}(h^{**})^2(s)\,ds\right)^{1/2}} \geq \frac{1}{2t}. \tag{9.29}$$

Under the assumptions (B), defining the function $h(s):[0.\pi]\to\mathbb{R}$ as above, we have

$$\lambda_2\left(S_{D^*}(t)\right) = \frac{1}{t}\frac{\left(\displaystyle\int\limits_0^{\pi}|h'(s)|^2\,ds\right)^{1/2}}{\left(\displaystyle\int\limits_0^{\pi}(h(s)-C)^2\,ds\right)^{1/2}}.$$

We remark that

$$\int\limits_0^{\pi}(h(s)-C)\,ds = 0$$

and denote by $\tilde{h}(s)$ the function $h(s)-C$. Extend by the symmetry the function $\tilde{h}(s)$ setting

$$\tilde{h}^*(s) = \tilde{h}(-s), \quad s\in[-\pi,0), \quad \tilde{h}^*(s)\Big|_{[0,\pi]} = \tilde{h}(s).$$

The function $\tilde{h}^*(s)$ satisfies the Lipschitz condition on $[-\pi,\pi]$ and has equal values on its ends. Moreover,

$$\int\limits_{-\pi}^{\pi}\tilde{h}^*(s)\,ds = 0.$$

Therefore, this function is extendable by the periodicity on all axis up to the 2π-periodic function $\tilde{h}^{**}(s)$ with the zero integral along the period. By the Wirtinger inequality, we obtain

$$\lambda_2\left(S_{D^*}(t)\right) = \frac{1}{t}\frac{\left(\displaystyle\int\limits_0^{2\pi}|(\tilde{h}^{**})'(s)|^2\,ds\right)^{1/2}}{\left(\displaystyle\int\limits_0^{2\pi}(\tilde{h}^{**})^2(s)\,ds\right)^{1/2}} \geq \frac{1}{t}. \tag{9.30}$$

Combining relations (9.22), (9.25), (9.28), (9.29), (9.30), we arrive to (9.18). □

9.4.2 A Conjugate Function

Let $D \subset \mathbb{R}^2$ be a simply connected domain and let x_0 be a fixed point. If $u(x)$ is a generalized solution of $L[u] = 0$ in D, then we can consider a conjugate function

$$v(x) = \int_{x_0}^{x} -A_2(x, \nabla u(x)) \, dx_1 + A_1(x, \nabla u(x)) \, dx_2 \, .$$

In case that the vector function $A : D \times \mathbb{R}^2 \to \mathbb{R}^2$ belongs to $C^1(D \times \mathbb{R}^2)$ and $u \in C^2(D)$, the conjugate function $v(x)$ is defined and univalent in D. In the general case, some additional verifications are necessary.

We consider a complex value function

$$w = u(x) + \frac{i}{\nu_1} \, v(x) \, .$$

Theorem 9.31 *A map $w : D \to \mathbb{R}^2$ is a quasi-regular with respect to the metric ds_Ω^2 with a distortion coefficient which does not exceed*

$$q(\nu) = \frac{1}{2} \left(1 + \nu^2 + \sqrt{\nu^4 + 2\nu^3 - 3} \right) . \tag{9.32}$$

In particular, there exist a homeomorphism $T : D \to \mathbb{R}^2$ and a holomorphic function $\Phi : T(D) \to \mathbb{R}^2$ such that $w(x) = \Phi \circ T(x)$.

Proof. By the assumption (9.2) for the vector function $A(x, \xi)$, the function v and, hence, the vector function w are locally Lipschiz. By Rademacher's theorem the vector function w are differentiable a.e. in D. For the proof of our statement, it is sufficient to check that at every point $x \in D$ where w is differentiable, the Jacobian of the map is not negative and

$$\max_{ds_\Omega^2 = 1} |dw(x)| \leq q(\nu) \min_{ds_\Omega^2 = 1} |dw(x)| \, . \tag{9.33}$$

Without loss of generality, we can assume that $\nu_1 = 1$ and $\nu_2 = \nu$. Then the Jacobian of the map w has the form

$$u_{x_1} v_{x_2} - u_{x_2} v_{x_1} = \sum_{i=1}^{2} u_{x_i} A_i(x, \nabla u) \, ,$$

and by the condition (9.1) it is not negative. Further, we have

$$|dw|^2 = (u_{x_1}^2 + A_2^2)\, dx_1^2 + 2\,(u_{x_1} u_{x_2} - A_1\, A_2) + (u_{x_2}^2 + A_1^2)\, dx_2^2$$

where

$$A_i = A_i(x, \nabla u) \quad (i = 1, 2)\,.$$

Consider the bundle of quadratic forms $|dw|^2 - \lambda\, ds_\Omega^2$. Let λ_1 be the minimal and λ_2 be the maximal characteristic numbers of this bundle. Then

$$\lambda_1 = \min_{ds_\Omega^2 = 1} |dw(x)|^2\,, \quad \lambda_2 = \max_{ds_\Omega^2 = 1} |dw(x)|^2\,.$$

Find the ratio λ_2/λ_1. The characteristic equation for the bundle is written in the form

$$\begin{vmatrix} (u_{x_1}^2 + A_2^2) - \lambda\,(1 + u_{x_1}^2) & (u_{x_1} u_{x_2} - A_1\, A_2) - \lambda\, u_{x_1}\, u_{x_2} \\[2mm] (u_{x_1} u_{x_2} - A_1\, A_2) - \lambda\, u_{x_1}\, u_{x_2} & (u_{x_2}^2 + A_1^2) - \lambda\,(1 + u_{x_2}^2) \end{vmatrix} = 0\,,$$

or

$$\lambda^2\,(1 + |\nabla u|^2) - \lambda\,(|\nabla u|^2 + |A|^2 + \langle \nabla u, A \rangle^2) + \langle \nabla u, A \rangle^2 = 0\,.$$

From here, setting

$$\mu = \frac{|\nabla u|^2 + |A|^2 + \langle \nabla u, A \rangle^2}{2\sqrt{1 + |\nabla u|^2}\langle \nabla u, A \rangle}\,,$$

we find

$$\left(\frac{\lambda_2}{\lambda_1}\right)^{1/2} = \mu + \sqrt{\mu^2 - 1}\,.$$

Assumptions (9.1), (9.2) for L imply relations

$$|\nabla u|^2 + |A|^2 + \langle \nabla u, A \rangle^2 \le (1 + \nu^2)\,|\nabla u|^2$$

and

$$|\nabla u|^2 \le \sqrt{1 + |\nabla u|^2}\,\langle \nabla u, A \rangle\,.$$

Therefore,

$$\mu \le \frac{1}{2}(1 + \nu^2)$$

whence the necessary estimate for the distortion coefficient follows.

To complete the proof of the theorem, we remark that by Theorem 3.8 we can introduce isothermal coordinates $\xi = (\xi_1, \xi_2)$ on the surface Ω, and there exists a homeomorphic map $h : D \to \mathbb{R}^2$ such that

$$(d\xi_1)^2 + (d\xi_2)^2 = ds_\Omega^2 \circ h^{-1}\,.$$

The function $w \circ h^{-1}$ is holomorphic with respect to the metric $|d\xi|^2$ and we can put $\Phi = w \circ h^{-1}$, $T = h$. For details, for example, see the monograph of Vekua [189, Chapter 2, §6].   □

9.4.3 The Growth of a Conjugate Function (I)

Suppose, as above, that for every $t > 0$ the second derivative $w''(\xi)$ is bounded in $D^*(t)$. In particular, from here it follows that the first derivative $w'(\xi)$ belongs to $\mathrm{Lip}\,(\overline{D^*(t)})$ in every half-disc $D^*(t)$ and functions $w(\xi)$, $w'(\xi)$ are continuously extendable up to the boundary ∂D^*.

Theorem 9.34 *Let $w(\xi)$ be a holomorphic function in the half-plane D^*. Suppose that $w(\xi)$ is different from the identical constant, has the bounded second derivative $w''(\xi)$ in every subdomain $D^*(t)$, $t > 0$, and that everywhere on ∂D^* its real part $\mathrm{Re}\,w(\xi) \equiv 0$. Then*

$$\varliminf_{t\to\infty} t^{-1}\mathrm{osc}\,(\mathrm{Im}\,w(\xi), |\xi| = t) > 0. \tag{9.35}$$

Proof. We fix arbitrarily $r > 0$. We put

$$g(\tau) = \begin{cases} 1 & \text{for} \quad \tau \le r, \\[2mm] \ln\frac{2r}{\tau}\big/\ln 2 & \text{for} \quad r < \tau < 2r, \\[2mm] 0 & \text{for} \quad \tau \ge 2r. \end{cases}$$

For a constant C, the vector field

$$(v(\xi) - C)\,g^2(|\xi|)\nabla u(\xi)$$

satisfies the Lipschitz condition in $\overline{D^*}$, and its support is contained in $\overline{D^*(2r)}$. Since $du = 0$ along ∂D^*, then by Green's formula, we have

$$\int_{\partial D^*} (v - C)\,g^2\,du = 2\iint_{D^*} (v - C)\,g\,du \wedge dg - \iint_{D^*} g^2\,du \wedge dv$$

where the symbol $\wedge$ means the exterior product.

From here, using Cauchy-Riemann conditions for the vector function $w = (u, v)$, we deduce

$$\iint_{D^*(2r)} g^2(|\xi|)\,|\nabla u(\xi)|^2\,d\xi_1 d\xi_2 \le 2\iint_{D^*(2r)} |v - C|\,|g|\,|\nabla g|\,|\nabla u|\,d\xi_1 d\xi_2.$$

Further,

$$\iint\limits_{D^*(2r)} g^2(|\xi|)\, |\nabla u(\xi)|^2\, d\xi_1 d\xi_2 \le$$

$$\le 2 \left(\iint\limits_{D^*(2r)} g^2(|\xi|)\, |\nabla u(\xi)|^2\, d\xi_1 d\xi_2 \iint\limits_{D^*(2r)} |v - C|^2\, |g|\, |\nabla g|^2\, d\xi_1 d\xi_2 \right)^{\frac{1}{2}},$$

and observing that $|g| \le 1$, we find

$$\iint\limits_{D^*(r)} |\nabla u(\xi)|^2\, d\xi_1 d\xi_2 \;\le\; \frac{4}{\ln^2 2} \iint\limits_{D^*(2r)\setminus D^*(r)} |v(\xi) - C|^2\, \frac{d\xi_1 d\xi_2}{|\xi|^2} \le$$

$$\le \frac{4\pi}{\ln^2 2} \max_{\xi \in D^*(2r)} |v(\xi) - C|^2 .$$

It is not difficult to check that by the maximum-minimum principle,

$$\operatorname{osc}\{\operatorname{Im}\, w,\, D^*(2r)\} = \operatorname{osc}\{\operatorname{Im}\, w,\, S_{D^*}(2r)\}.$$

Thus we arrive to the relation

$$\int\limits_{D^*(r)} |\nabla u(\xi)|^2\, d\xi_1 d\xi_2 \le \frac{4\pi}{\ln^2 2} \operatorname{osc}^2\{\operatorname{Im}\, w,\, S_{D^*}(2r)\}. \tag{9.36}$$

Now we use Theorem 9.16. By (9.36) for arbitrary $0 < r < R < \infty$, it is fulfilled

$$I(r) \le \frac{4\pi}{\ln^2 2} \operatorname{osc}^2\{\operatorname{Im}\, w,\, S_{D^*}(2R)\} \left(\frac{r}{R} \right)^{1/2}.$$

From here we conclude that under a violation of (9.35), the integral $I(r)$ vanishes. Consequently, $w \equiv \operatorname{const}$ in $D^*(r)$, and that is impossible. □

9.4.4 The Growth of a Conjugate Function (II)

We put

$$\alpha(t) = \iint\limits_{D(1,t)} \frac{dx_1 dx_2}{|x|^2} + 2\nu\pi^2 .$$

The following statement imposes some conditions to the growth speed of a conjugate function $v(x)$ in a simply connected domain of the general form and to solutions of minimal surface type equations.

Theorem 9.37 *Let $D \subset \mathbb{R}^2$ be an elementary domain. Let u be a generalized solution of the equation (9.3) with (9.1), (9.2), and let*

$$u \in \operatorname{Lip} \overline{D(t)} \quad \text{for every} \quad t > 0.$$

Suppose that everywhere on the boundary ∂D, it is fulfilled $u \equiv 0$. If the function $v(x)$ is different from the identical constant in D, then for every $\varepsilon \in (0,1)$, the following relation holds

$$\underline{\lim}_{t\to\infty} \operatorname{osc}\{v(x), S_D(t)\} \exp\left\{-\frac{2(1-\varepsilon)}{\nu\, q(\nu)} \frac{\ln^2 t}{\alpha(t)}\right\} > 0. \tag{9.38}$$

Here

$$\alpha(t) = \iint_{D(1,t)} \frac{dx_1 dx_2}{|x|^2} + 2\nu\pi^2.$$

Proof. We use the auxiliary map $\xi = \xi(x)$ and pass to isothermal coordinates on the graph Ω of $x_3 = u(x_1, x_2)$. Since $\xi(x) : D \to D^*$ transforms ellipses $ds_\Omega^2 = 1$ onto circles, then the function $w^* = w(x(\xi))$ has in the half-plane D^* the distortion coefficient with respect to the Euclidean metric which does not exceed $q = q(\nu)$. The real part of $w^*(\xi)$ is identical constant on the boundary ∂D^*.

Lemma 9.39 *If $\operatorname{Im} w^*(\xi)$ is different from the identical constant in the half-plane D^*, then*

$$\underline{\lim}_{\tau\to\infty} \tau^{-1/q} \operatorname{osc}\{\operatorname{Im} w^*(\xi), |\xi| = \tau\} > 0. \tag{9.40}$$

The **proof** is based on the following arguments. The function $w^*(\xi)$ is representable in the form $\Phi \circ T$ where Φ is a holomorphic function and $T : D^* \to D^*$ is a schlicht $q(\nu)$-quasi-conformal map of the half-plane onto itself normed by conditions

$$T(0,0) = (0,0), \quad T(1,0) = (1,0), \quad T(\infty) = \infty.$$

Extending the map T by the symmetry onto the plane $\mathbb{R}^2$ and using Theorem 3.49, we conclude that for sufficiently big $\xi \in D^*$, it is fulfilled

$$C_1 |\xi|^{1/q(\nu)} \le |T(\xi)| \le C_2 |\xi|^{q(\nu)}$$

where C_1, C_2 are some constants. The relation (9.40) is now the direct corollary of Theorem 9.34. $\square$

The **proof of Theorem 9.37** completes with the reference to Theorem 9.8. Indeed, by the maximum principle for every $t > 0$, we have

$$\operatorname{osc}\{v(x),\, S_D(t)\} \geq \operatorname{osc}\{\operatorname{Im} w^*(\xi),\, |\xi| = m(t)\}.$$

From here the following inequality holds

$$\underline{\lim}_{t\to\infty} m^{-1/q(\nu)}(t)\operatorname{osc}\{v(x),\, S_D(t)\} \geq$$

$$\geq \underline{\lim}_{t\to\infty} m^{-1/q(\nu)}(t)\operatorname{osc}\{\operatorname{Im} w^*(\xi),\, |\xi| = m(t)\},$$

and by (9.40)

$$\underline{\lim}_{t\to\infty} m^{-1/q(\nu)}(t)\operatorname{osc}\{v(x),\, S_D(t)\} > 0.$$

On the other hand, from (9.9) it follows that for every $\varepsilon > 0$ for sufficiently big $t > 0$, we have

$$m(t) \geq \exp\left\{\frac{2(1-\varepsilon)}{\nu}\frac{\ln^2 t}{\alpha(t)}\right\}.$$

Combining found relations, we arrive to (9.38). $\square$

9.5   Narrow Domains

Let us agree on designations. We say that a simply connected domain $\mathcal{D} \subset \mathbb{R}^2$ is m-domain if it has equally $m < \infty$ different components of the boundary $\partial\mathcal{D}$. In particular, a 0-domain is the all plane $\mathbb{R}^2$.

We put

$$\beta(\mathcal{D}) = \underline{\lim}_{t\to\infty}\frac{1}{\ln t}\iint\limits_{\mathcal{D}(1,t)}\frac{dx_1 dx_2}{|x|^2}.$$

The quantity $\beta(\mathcal{D})$ is a characteristic of the narrowness of the domain $\mathcal{D}$ in a neighborhood of the infinite point of $\mathbb{R}^2$. In the case if $\mathcal{D}$ is a corner of angle θ, we have $\beta(\mathcal{D}) = \theta$.

If $\mathcal{D}_1$, $\mathcal{D}_2$, ..., $\mathcal{D}_N$ is a system of non-overlapping domains lying in $\mathcal{D}$, then it is easy to see that

$$\sum_{i=1}^{N}\beta(\mathcal{D}_i) \leq \beta(\mathcal{D}). \tag{9.41}$$

The following analog of Nitsche's theorem holds in simply connected domains.

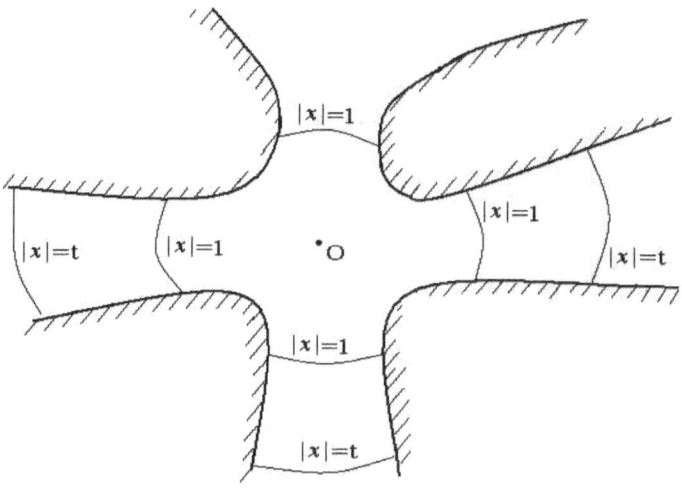

Fig. 9.1.

Theorem 9.42   *Let $\mathcal{D}$ be an arbitrary elementary m-domain in $\mathbb{R}^2$, $m \geq 1$, and let $u(x)$ be a solution of the minimal surface type equation (9.1), (9.2), (9.3), satisfying $u|_{\partial \mathcal{D}} = 0$. If*

$$\beta(\mathcal{D}) < \frac{2}{\nu\, q(\nu)}\,,$$

then $u(x) \equiv 0$ everywhere in $\mathcal{D}$.

We will need the Phragmén-Lindelöf principle for generalized solutions.

Lemma 9.43   *Let $D \subset \mathbb{R}^2$ be an unbounded domain, $\partial D \neq \emptyset$. If $u(x)$ is a generalized solution of the minimal surface type equation (9.1), (9.2), (9.3) in D, satisfying the condition $u|_{\partial D} = 0$, then either $u \equiv 0$ in D or*

$$\varliminf_{r \to \infty} \frac{M_u(r)}{\ln^{1/2} r} > 0\,, \quad M_u(r) = \max_{x \in S_D(r)} |u(x)|\,.$$

Proof.   Let $0 < r < R < \infty$ be fixed numbers. Choose the function

$\psi : (0, +\infty) \to (0, +\infty)$ in the form

$$\psi(t) = \begin{cases} 1 & \text{for} \quad 0 < t \le r\,, \\[2mm] \dfrac{\ln t/r}{\ln R/r} & \text{for} \quad r < t < R\,, \\[2mm] 0 & \text{for} \quad R \le t < +\infty\,. \end{cases}$$

The function $\varphi(x) = u(x)\,\psi^2(|x|)$ has the compact support

$$\mathrm{supp}_{\tilde{D}}\varphi \subset \tilde{D} \cap \{x : |x| \le R\}\,.$$

By the definition (9.5) of the generalized solution, we can write

$$\iint\limits_{D} \langle \nabla(u\,\psi^2), A(x, \nabla u)\rangle \, dx_1 \, dx_2 = 0\,.$$

From here

$$\iint\limits_{D} \psi^2 \langle \nabla u, A(x, \nabla u)\rangle \, dx_1 \, dx_2 = -2 \iint\limits_{D} u\,\psi \, \langle \nabla \psi, A(x, \nabla u)\rangle \, dx_1 \, dx_2\,,$$

and further, by the Cauchy integral inequality we have

$$\iint\limits_{D} \psi^2 \langle \nabla u, A(x, \nabla u)\rangle \, dx_1 \, dx_2 \le$$

$$\le 2 \left(\iint\limits_{D} \psi^2 \, |A(x, \nabla u)|^2 \, dx_1 \, dx_2 \right)^{1/2} \left(\iint\limits_{D} u^2 \, |\nabla \psi|^2 \, dx_1 \, dx_2 \right)^{1/2}\,.$$

From structure conditions (9.1) and (9.2), it follows

$$|A(x, \nabla f)|^2 \le \frac{\nu_2^2}{\nu_1} \langle \nabla u, A(x, \nabla u)\rangle\,,$$

we find

$$\iint\limits_{D} \psi^2 \langle \nabla u, A(x, \nabla u)\rangle \, dx_1 \, dx_2 \le 4 \frac{\nu_2^2}{\nu_1} M_u^2(R) \iint\limits_{D} |\nabla \psi|^2 \, dx_1 \, dx_2\,.$$

Thus using the special choice of the function ψ, we obtain

$$\iint\limits_{B_D(r)} \langle \nabla u, A(x, \nabla u) \rangle \, dx_1 \, dx_2 \le 4 \frac{\nu_2^2}{\nu_1} \frac{M_u^2(R)}{\ln^2 R/r} \iint\limits_{r < |x| < R} \frac{dx_1 \, dx_2}{|x|^2}$$

or,

$$\iint\limits_{B_D(r)} \langle \nabla u, A(x, \nabla u) \rangle \, dx_1 \, dx_2 \le 8\pi \frac{\nu_2^2}{\nu_1} \frac{M_u^2(R)}{\ln R/r} .$$

If $M_U(R)$ grows more slowly than it is assumed in the lemma, then by (9.1) we can conclude that $\nabla u \equiv 0$ in D. Therefore, $u \equiv 0$ in D. □

Pass to the **proof of Theorem 9.42**. Because the equation satisfies assumptions (9.1), (9.2), then by Lemma 9.43 the solution $u(x)$ is an identical constant or unbounded in $\mathcal{D}$. Suppose that at a point $x_0 \in \mathcal{D}$ the solution is positive. Denote by K_i $(i = 1, 2, \ldots, m)$ components of the boundary $\partial \mathcal{D}$, by d_i – the distance from x_0 to K_i. Let

$$d = \max_{1 \le i \le m} d_i .$$

Consider the set O_t where

$$u(x) > t \max_{|x - x_0| \le d} u(x) .$$

Since $u(x)$ is locally Lipschitz, then for a.e. $t > 0$ the boundary ∂O_t is locally rectifiable [37, Theorem **3.2.31**].

Choose $t > 1$ with the previously mentioned property. Let Σ be a component of the set ∂O_t. The set Σ divides the domain $\mathcal{D}$ into two subdomains $\mathcal{D}_1$ and $\mathcal{D}_2$ every of which is unbounded by the maximum principle. Because Σ can not intersect the disc $|x - x_0| \le d$, one of the domains, for example $\mathcal{D}_1$, is elementary.

On the boundary of $\mathcal{D}_1$, the solution $u(x)$ is constant. Moreover,

$$\beta(\mathcal{D}_1) \le \beta(\mathcal{D}) < \frac{2}{\nu \, q(\nu)} .$$

From here it follows that $u(x) \equiv$ const on $\mathcal{D}_1$.

To check this statement, we use Theorem 9.37. Choose $0 < \varepsilon_1 < 1$ so small that for a sequence $t_k \to \infty$:

$$\frac{\alpha(t_k)}{\ln t_k} \le \frac{2(1 - \varepsilon_1)}{\nu \, q(\nu)} .$$

Then by (9.38) we can write

$$\underline{\lim}_{k\to\infty} t^{-(1-\varepsilon)/(1-\varepsilon_1)} \operatorname{osc}\{v(x), S_D(t_k)\} > 0 \quad (\forall\, \varepsilon > 0)\,,$$

and, setting $\varepsilon < \varepsilon_1$, we find a sequence of arcs $S_D(t_k)$ where the oscillation of $v(x)$ grows more rapid than linear functions. However, by the assumption (9.2)

$$|\nabla v(x)| \le |A(x, \nabla u(x))| \le \nu_2$$

and

$$\operatorname{osc}\{v(x), S_D(t_k)\} \le 2\pi\nu_2\, t_k\,.$$

The obtained contradiction permits to conclude that $v(x) \equiv \text{const}$ in $\mathcal{D}_1$ and $u(x) \equiv 0$ in $\mathcal{D}$. $\qquad\qquad\qquad\qquad\qquad\qquad\qquad\qquad\qquad\qquad$ □

Remark 9.44   *If* $L[u] = 0$ *is the minimal surface equation, then constants* ν *and* $q(\nu)$ *are equal to 1. In this case, if* $\mathcal{D}$ *is a corner of angle* α, *then the statement of Theorem 9.42 is valid only for* $\alpha < 2$. *It is a little worse than Nitsche's result and it is desirable to prove Theorem 9.42 for* $\alpha < \pi$.

This wish was formulated in our article [113] (1981). A close problem was set up by Meeks in his lecture "The Global Theory of Minimal Surfaces" in Clay Mathematics Institute Summer School (2001). Meeks conjectured that there can be at most two solutions with zero boundary data over disjoint subdomains of the plane. Spruck [173] proved Meeks' supposition under some special conditions to the solution $u(x)$. Previously, Li and Wang [96] proved that the number of such domains can not exceed 12. Theorem 9.42 implies that this number is not greater than 3.

The analogous problem for solutions of *maximal surface type equations* is considered in [116], [69, §**3.7**].

From Theorem 9.42, we obtain the following statement about the uniqueness in the Dirichlét problem for the minimal surface type equation.

Corollary 9.45   *Let* $\mathcal{D}$ *be an* m-*domain,* $m \ge 1$, *and let* $\varphi(x)$ *be a continuous bounded function defined everywhere on the boundary* $\partial\mathcal{D}$. *Let* $u_i(x) : \overline{\mathcal{D}} \to \mathbb{R}$ $(i = 1, 2)$ *be continuous solutions of a minimal surface type equation* (9.1), (9.2), (9.3) *such that*

$$u_i(x)|_{\partial\mathcal{D}} = \varphi(x) \quad (i = 1, 2)\,.$$

If $\beta(\mathcal{D}) < 2/(\nu\, q(\nu))$, *then* $u_1(x) = u_2(x)$ *everywhere in the domain* $\mathcal{D}$.

The **proof** will follow immediately after Lemma 9.43 (see also [111, Corollary **4.1**]) if we can prove that every solution $u_i(x)$ $(i = 1, 2)$ is bounded in $\mathcal{D}$.

Suppose the contrary. Without loss of generality, we can assume that the function $u_i(x)$ is upper unbounded in $\mathcal{D}$. Then there exist constants $c_1 > c > 0$ such that $\varphi(x) < c$ everywhere on $\partial\mathcal{D}$, and the set

$$O = \{x \in \mathcal{D} : u_i(x) > c_1\}$$

is not empty. By the maximum principle, ∂O has no compact components. Fix a point $x_0 \in O$ and determine by K_i $(i = 1, 2, \ldots, m)$ components of ∂O separating x_0 from the corresponding component of $\partial\mathcal{D}$. Consider a subdomain $\mathcal{D}'$ of $\mathcal{D}$, containing x_0 and having the set $\cap_{i=1}^m K_i$ as the boundary. The domain $\mathcal{D}'$ is m-domain. Since $u_i(x) = c_1$ on $\partial\mathcal{D}'$ and

$$\beta(\mathcal{D}') \leq \beta(\mathcal{D}) < 2/(\nu q(\nu)),$$

then by Theorem 9.42 we conclude that $u_i(x) \equiv c_1$ in $\mathcal{D}'$, what is impossible. $\square$

There is another application of the theorem about narrow domains. We joint it with a condition of constant sign solutions.

Corollary 9.46 *Let $\mathcal{D}$ be an m-domain, $m \geq 1$, and let $u(x)$ be a continuous in $\overline{\mathcal{D}}$ solution of a minimal surface type equation (9.1), (9.2), (9.3) such that $u(x)|_{\partial\mathcal{D}} = 0$. If $\beta(\mathcal{D}) < 4/(\nu\, q(\nu))$, then $u(x)$ does not change its sign in $\mathcal{D}$.*

Proof. Indeed, suppose that a solution changes the sign. Considering as above, we can find two non-overlapping subdomains $\mathcal{D}_1$, $\mathcal{D}_2$ of $\mathcal{D}$ on which boundaries the solution $u(x)$ is constant. Using the inequality (9.41), we have

$$\beta(\mathcal{D}_1) + \beta(\mathcal{D}_2) < \frac{4}{\nu\, q(\nu)},$$

and at least one domain satisfies the assumptions of Theorem 9.42. Thus, at least in one domain, $u(x)$ is constant, what is impossible. $\square$

9.6 Critical Points of a Solution

By Morse [131, Chapter I, §2], we determine the concept of the index of a critical point of $u(x)$. First of all we remark that by Theorem 9.31 the map $w(x) = u(x) + \dfrac{i}{\nu_1} v(x)$ is a composition of a holomorphic function and a homeomorphism, i.e. $w(x)$ is *pseudoholomorphic* (or *pseudoanalytic*) function[2]. Thus its real and imaginary parts are *pseudoharmonic* functions, i.e. – representable in the form of superpositions $h \circ T$ of suitable harmonic functions h and homeomorphisms T.

Let $x_0 \in \mathcal{D}$ be a point and let E be the set where $u(x) = u(x_0)$. The set E divides a neighborhood of x_0 into $2s$ curvilinear sectors. The point x_0 is called *critical* point of a solution $u(x)$ if $s > 1$. The number $s - 1$ is called the *topological index* of $u(x)$ at x_0 and denoted with $i(x_0)$.

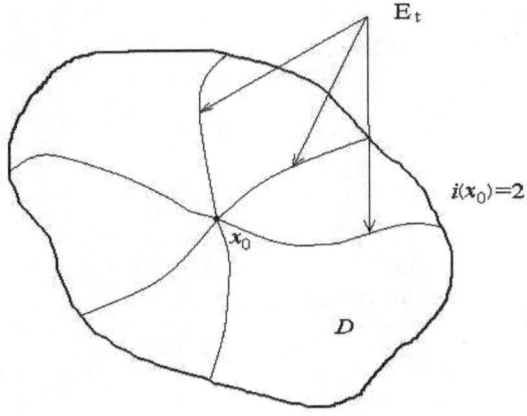

Fig. 9.2.

Since $u(x)$ is pseudoharmonic, then its critical points are isolated. For harmonic functions in subdomains of $\mathbb{R}^2$, it is the unique restriction onto the distribution of critical points. From the Weierstrass theorem about representation with the infinite product of holomorphic functions, it follows that every sequence of points $a_1, a_2, \ldots$, having no limiting points in the domain, can be a sequence of critical points of a harmonic function defined in this domain (see, for example, [166, p. 235]). The analogous property can be shown for solutions of linear partial elliptic equations of the sufficiently general form. Assumptions (9.1), (9.2) for the minimal surface type equation (9.3) imply essential restrictions onto a number and indices of critical points of the solution $u(x)$.

Namely, the following theorem holds.

---

[2]About pseudoanalitic functions, see Stoilow [174, Chapter V], Avkhadiev [10, Chapter 1].

Theorem 9.47 *Let $\mathcal{D}$ be an elementary m-domain with the boundary $\partial\mathcal{D}$ consisted of simple Jordan arcs. If $m > 0$, then let $\varphi(x)$ be a continuous function, defined on $\partial\mathcal{D}$ and having there $j,\ 0 \le j < \infty$, points of the local extremum. Suppose that $u(x)$ is continuous in $\overline{\mathcal{D}}$ solution of a minimal surface type equation (9.1), (9.2), (9.3), satisfying to*

$$u(x)|_{\partial\mathcal{D}} = \varphi(x).$$

Then $u(x)$ has in $\mathcal{D}$ a finite number of critical points $a_1, a_2, \ldots, a_N$ such that

$$2\sum_{k=1}^{N} i(a_k) - \delta(m,j) + 2 \le \frac{1}{2}\nu q(\nu)\,\beta(\mathcal{D}) \tag{9.48}$$

where $\delta(m,j) = 2m + j$ in the general case and $\delta(m,0) = m$ for $\varphi \equiv 0$ on $\partial\mathcal{D}$.

Proof. Let $a_1, a_2, \ldots, a_k, \ldots$ be critical points of the solution $u(x)$. Denote by E_k the component of the set

$$\{x \in \mathcal{D} : u(x) = u(a_k)\}$$

which contains a_k.

Consider the set E_1. Because $\mathcal{D}$ is simply connected, then by the maximum principle, every component $O_{1,n}$ of $\mathcal{D}\setminus E_1$ is simply connected and unbounded. Critical points of the solution $u(x)$ are isolated, and hence, a sufficient small neighborhood of the point a_1 is divided with the set E_1 exactly into

$$l_1 = 2\,i(a_1) + 2$$

curvilinear sectors. Let $\Delta_{11}, \ldots, \Delta_{1l_1}$ be domains of the run $\{O_{1,n}\}$ which contains a_1 in its closuring. The number of such domains is not less than l_1, since E_1 can not have closed cycles.

Consider E_2. Suppose that $E_1 \cap E_2 \ne \emptyset$. Then $E_1 = E_2$, and there exist $l_2 = 2\,i(a_2)$ components $\Delta_{21}, \ldots, \Delta_{2l_2}$ of $\mathcal{D}\setminus E_1$, abutting to the point a_2 and different from Δ_{1n}. Thus we find

$$2\,i(a_1) + 2\,i(a_2) + 2$$

non-overlapping, unbounded, simply connected domains, lying in $\mathcal{D}$ and having on its boundaries constant values of $u(x)$.

Let $E_1 \cap E_2 = \emptyset$ and let the point a_2 belongs to a domain of the run $\{\Delta_{1n}\}$, for example, to Δ_{11}. The set Δ_{11} is divided with the continuum E_2 into domains $O_{2,n}$ from which exactly

$$l_2 = 2\,i(a_2) + 1$$

domains Δ_{21}, ..., Δ_{2l_2} are abutting to a_2, simply connected and unbounded. In this case also, there are

$$2\,i(a_1) + 2\,i(a_2) + 2$$

non-overlapping domains in $\mathcal{D}$ on boundaries of which $u(x)$ is constant.

Analogously, we analyze the case in which $E_1 \cap E_2 = \emptyset$ but a_2 does not belong to $\{\Delta_{1n}\}$.

Further, we consider in turn E_3, ..., E_N. After N steps we find at least

$$2s = 2\sum_{k=1}^{N} i(a_k) + 2$$

non-overlapping subdomains of $\mathcal{D}$ with described properties. Some number of these domains abut to the boundary $\partial\mathcal{D}$. In the general case, we have no more than $2m + j$ such domains, and there exist at least

$$p = 2\,s - 2\,n - j$$

non-overlapping unbounded domains $\mathcal{D}_1$, ..., $\mathcal{D}_p$ on boundaries of which $u(x)$ is constant.

From (9.41) it follows that at least one among them, for example $\mathcal{D}_1$, has the property

$$\beta(\mathcal{D}_1) \le \frac{1}{p}\,\beta(\mathcal{D})\,.$$

Thus, if the inequality (9.48) is not valid, we have

$$\frac{1}{2}\,p\,\nu\,q(\nu)\,\beta(\mathcal{D}_1) \le \frac{1}{2}\,\nu\,q(\nu)\,\beta(\mathcal{D}) < 2s - 2m - j = p\,,$$

and hence,

$$\beta(\mathcal{D}_1) < \frac{2}{\nu\,q(\nu)}\,.$$

From here by Theorem 9.42, we conclude that $u(x) \equiv \text{const}$ in $\mathcal{D}_1$. That is impossible.

In the case $\varphi(x) \equiv 0$ on ∂D, it is necessary to exclude from the consideration only subdomains, containing components of ∂D as components of those boundaries. The number of such domains does not exceed m. Further reasonings are the same as before, and the prove of the theorem is completed. □

Corollary 9.49 *If $u(x)$ is an entire solution of (9.3) with structure conditions (9.1), (9.2), and*

$$\nu\,q(\nu) < \frac{4}{\pi}\,,$$

then $u(x)$ has no critical points.

Corollary 9.50   *Let $\mathcal{D}$ be a 1-domain and let $u(x)$ be a solution of (9.3) vanishing on $\partial\mathcal{D}$. Suppose that the equation satisfies to (9.1), (9.2), and*

$$\beta(\mathcal{D}) \leq \frac{6}{\nu\, q(\nu)}\,.$$

Then $u(x)$ has no critical points in $\mathcal{D}$.

For solutions of the minimal surface type equation (2.20), the last condition is rewritten in the form $\beta(\mathcal{D}) < 6$. Assuming that Theorem 9.42 is valid for $\beta(\mathcal{D}) < \pi$, we can conclude as above that solutions of (2.20) which are zero on the boundary $\partial\mathcal{D}$ can not have critical points in 1-domain.

The example of the function

$$u(x_1, x_2) = \sqrt{\operatorname{ch}^2 x_1 - x_2^2} \quad \text{(upper one-half of a catenoid)}$$

shows that, already in 2-domains, there exist solutions of (2.20), having critical points. This function is defined in the 2-domain $\mathcal{D}$, described with the inequality

$$\{x = (x_1, x_2) \in \mathbb{R}^2 : x_2^2 < \operatorname{ch}^2 x_1\},$$

equals zero on $\partial\mathcal{D}$ and has the critical point $(0,0)$ of the index 1.

Open questions 9.51   1) Describe all domains where the maximum principle for solutions of (2.20) is valid in the previously mentioned reinforced form.   2) Find estimates of the summary topological index for entire solutions of (2.20) which are better than in Theorem 9.47.   3) Widen the class of equations (2.20), for which previously mentioned effects are valid.

Chapter 10

Solutions Close to a Boundary

We study generalized solutions of minimal surface type equation. We prove that every solution has on the boundary no more than a countable number of jumps. In particular, every solution, defined in the disc exterior, is continuously extendable up to the boundary circle, excepting possibly a countable set of points. For Fatou's type theorems about angular boundary values, see [109], [83], [115], [100], [120].

10.1 Main Results

Let $D \subset \mathbb{R}^2$ be a domain and let $e \subset D$ be a set of zero linear Hausdorff measure.

Let
$$A = (A_1(x,\xi), A_2(x,\xi)) : (D \setminus e) \times \mathbb{R}^2 \to \mathbb{R}^2$$

be a continuous vector function. Suppose that, for every point $x = (x_1, x_2) \in D \setminus e$ and every $\xi = (\xi_1, \xi_2) \in \mathbb{R}^2$, the following structure conditions hold:

$$\nu_1 \frac{|\xi|^2}{\sqrt{1+|\xi|^2}} \leq \sum_{i=1}^{2} \xi_i \, A_i(x,\xi), \tag{10.1}$$

$$|A(x,\xi)| \leq \nu_2(x) \tag{10.2}$$

where ν_1 is a positive constant and $\nu_2(x)$ is a positive continuous function.

Below we will assume that $\nu_2(x)$ satisfies

$$\lim_{\{D_n\}} \int_{\partial D_n} \nu_2(x)\,|dx| = \nu_2^* < \infty \tag{10.3}$$

where the lower limit is taken over all sequences $\{D_n\}$ of subdomains of D with rectifiable boundaries for which $\overline{D_n} \subset D_{n+1}$, $\cup_{n=1}^\infty \overline{D_n} = D$. (Here and below, the symbol $\overline{H}$ means the closure of the set $H \subset \mathbb{R}^2$ with respect to the topology $\mathbb{R}^2$.)

Consider the equation

$$\sum_{i=1}^{2} \frac{d}{dx_i} A_i(x, \nabla f) = 0. \tag{10.4}$$

As above in Chapter 8, we use the following definition of the generalized solution. Denote by $D_b(f)$ the subset of D, at every point of which the function f is not differentiable. By a *generalized solution* of (10.4), we call a locally Lipschitz function f with the following property. For every bounded subdomain Δ, $\overline{\Delta} \subset D$, with the rectifiable boundary $\partial\Delta$, $\mathrm{mes}_1(\partial\Delta \cap D_b(f)) = 0$, and a function $\varphi \in \mathrm{Lip}\,\overline{\Delta}$, the following equality holds

$$\int_{\partial\Delta} \varphi \sum_{i=1}^{2} A_i(x, \nabla f)\, n_i\,|dx| = \iint_{\Delta} \sum_{i=1}^{2} \varphi'_{x_i} A_i(x, \nabla f)\, dx_1\, dx_2. \tag{10.5}$$

Here and below, $\overline{n} = (n_1(x), n_2(x))$ is a unit vector of the exterior normal to the boundary $\partial\Delta$.

The set of the discontinuity points of the vector function A has zero linear Hausdorff measure, and hence, the contour integral in (10.5) exists.

Exercise 10.6 Prove (or refute !) that classes of generalization solutions of minimal surface type equations (9.1), (9.2), (9.3), introduced here and in Chapter 9, coincide.

□

Let f be a continuous function, defined in the domain $D \subset \mathbb{R}^2$ with the rectifiable boundary. The function f has a *finite (or infinite) angular boundary value* α at $a = (a_1, a_2) \in \partial D$ if

$$\lim_{x \to a} f(x) = \alpha$$

along every angle C with a vertex at a lying inside D.

The following theorem is a version of Fatou's theorem [147, Chapter I, §5]. *Every bounded, harmonic in a unit disc B function f has angular boundary values a.e. on the circle* ∂B.

Moreover, there are examples of unbounded harmonic functions which have no angular boundary values on sets $H \subset \partial B$ with linear measure $\text{mes}_1 H > 0$ [28, Chapter 2, §10].

Problem. (J.C.C. Nitsche [137, Chapter VII, n. **4**]) Do valid Fatou's type theorems exist for solutions of (2.20)?

The following result was obtained in [115].

(A) *Every solution of the minimal surface equation* (2.20) *has finite or infinite angular boundary values a.e. on* ∂B.
Let's emphasize that here we do not suppose additional restrictions for solutions.

At the same time, it is necessary to remark that the behavior of solutions of (2.20) depends on the specific structure of the domain where these solutions are defined. Namely, for the solutions, defined in the exterior of a disc B, the following statement holds [115].

(B) *Every solution of the minimal surface equation* (2.20), *defined over* $\mathbb{R}^2 \setminus \overline{B}$, *is continuously extended a.e. to the boundary, i.e. a finite limit*

$$\lim_{x \to a} f(x) = \alpha, \quad x \in \mathbb{R}^2 \setminus \overline{B}$$

exists a.e. on the circle ∂B.

Theorem 10.7   *Let $D \subset \mathbb{R}^2$ be a domain with the Jordan rectifiable boundary. Every generalized solution of the equation* (10.4) *with the structure conditions* (10.1), (10.2), (10.3) *has finite or infinite angular boundary values a.e. on* ∂D.

On generalizations, see [100], [120].

Example 10.8   A solution of (2.20) can have infinite values on a set of the positive linear measure. Consider the Scherk surface

$$f(x_1, x_2) = \log \frac{\cos x_2}{\cos x_1},$$

defined over the square

$$Q = \left\{ (x_1, x_2) : -\frac{\pi}{2} < x_i < \frac{\pi}{2} \quad (i = 1, 2) \right\}.$$

This surface is minimal, and the function f equals $\pm\infty$ on horizontal and vertical sides of the square boundary. In the vertices of the square, the function $f|_{\partial Q}$ has jumps.

□

Let $D \subset \mathbb{R}^2$ be a simply connected domain with the Jordan boundary ∂D and let $O \in D$ be a fixed point. Let $U \subset D$ be an open set. We denote: $[U] = \overline{U} \setminus \partial D$, $\partial' U = [U] \setminus U$.

Definition 10.9 *Let f be a continuous function defined in D. We will call $a \in \partial D$ by the point of quasi-continuity f if there is a sequence of subdomains $\{D_k\}_{k=1}^{\infty}$ of the domain D with properties:*

(α)   *every $\partial' D_k$ separates a from the fixed point O;*

(β) $\cap_k [D_k] = \emptyset$;   *length $\partial' D_k \to 0$ for $k \to \infty$;*

(γ) $\liminf_{k \to \infty} \mathrm{osc}(\partial' D_k, f) = 0$.
We will call all remaining points of ∂D by jump points of the function f.

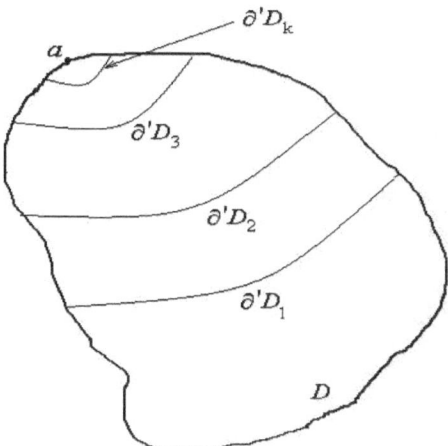

Fig. 10.1.

Define the following nonlocal characteristic of a boundary point $a \in \partial D$. We put

$$\delta(a, f, O) = \inf_{\gamma} \max\{\mathrm{osc}\,(\gamma, f), \mathrm{length}\,(\gamma)\}$$

where the infimum is taken over all arcs $\gamma \subset D$, $\overline{\gamma} \cap \partial D \neq \emptyset$, separating points a and O in D.

It is clear that f is quasi-continuous at a point $a \in \partial D$ if and only if $\delta(a, f, O) = 0$.

On the structure of solutions of (2.20) close to jump points, see Lancaster [83]. We prove that boundary quasi-continuity points of solutions f of the minimal surface type equation are typical.

Consider the set $\mathcal{H}$ of all piecewise continuous functions $h : \mathbb{R} \to \mathbb{R}$ with properties:

(i) $0 \le h(t) \le 1$ for all $t \in \mathbb{R}$;

(ii) $\int\limits_{-\infty}^{+\infty} h(t)\, dt \le h_0 < \infty, \quad h_0 = \text{const.}$

Theorem 10.10   *Let $D \subset \mathbb{R}^2$ be a simply connected domain bounded with a simple rectifiable Jordan curve ∂D, $O \in D$. Let $f : D \to \mathbb{R}$ be an arbitrary solution of an equation (10.4) with structure restrictions (10.1), (10.2), (10.3).*

Then for every $h \in \mathcal{H}$, the function

$$w(x) = \int\limits_{-\infty}^{f(x)} h(t)\, dt$$

is quasi-continuous at all points $a \in \partial D$, except only in a countable set.

Moreover, if $a_1, a_2, \ldots$ are jump points of f, then

$$\sum_{i=1}^{3} \max\left\{ \pi - \frac{K^2}{\pi \delta^2(a_i, w, O)}, 0 \right\} +$$

$$+ 2 \sum_{i=4}^{\infty} \arcsin\left(\tfrac{1}{2} \exp\left\{ -\frac{K^2}{\delta^2(a_i, w, O)} \right\} \right) \le \pi \tag{10.11}$$

where

$$K = 2\sqrt{\pi}\left(h_0 \frac{\nu_2^*}{\nu_1} + 2\, \mathrm{mes}_2\,(D) \right)^{1/2}.$$

Example 10.12   Suppose that $|f| < 1$ everywhere in D. We put

$$h(t) = \begin{cases} 1 & \text{for} \quad t \in [-1, 1], \\ 0 & \text{for} \quad |t| > 1. \end{cases}$$

For the auxiliary function

$$w(x) = \int\limits_{-\infty}^{f(x)} h(t)\, dt,$$

we have $0 \leq w(x) \leq 2$. Points of quasi-continuity of $f(x)$ are points of quasi-continuity of $w(x)$. Moreover,

$$\delta(a_k, w, O) = \delta(a_k, f, O) \quad \text{for all} \quad k = 1, 2, \dots .$$

□

Example 10.13 In the case of an unbounded function $f(x)$, we can put

$$h(t) = \frac{1}{1 + t^2}.$$

Then

$$w(x) = \int\limits_{-\infty}^{f(x)} \frac{dt}{1 + t^2} = \operatorname{arctg} f(x) + \frac{\pi}{2}.$$

Here Theorem 10.10 guarantees that $\operatorname{arctg} f(x)$ is quasi-continuous. It is clear that in the general case, the quasi-continuity of this function does not imply the quasi-continuity of $f(x)$. The function $f(x)$ from Example 10.8 is not quasi-continuous in every boundary point. However, $\operatorname{arctg} f(x)$ is quasi-continuous everywhere on the boundary of the square except its vertices.

□

As a corollary of Theorem 10.10, we have the following statement.

Theorem 10.14 Let $\mathcal{E} = \{x = (x_1, x_2) \in \mathbb{R}^2 : x_1^2 + x_2^2 > 1\}$ be an exterior of a unit circle C. Let $D \subset \mathcal{E}$ be a domain and $\Gamma \subset (C \cap \partial D)$ be a boundary arc.

Then every solution f of the minimal surface equation (2.20) in D is continuously extendable at an arbitrary point $a \in \Gamma$, except at most, in a countable set.

Open questions 10.15 1) Find analogs of Theorem 10.14 for solutions of minimal surface type equations. 2) Prove (or refute) Theorem 10.14 with the exchange in its formulation the boundary circle replaced by a concave curve of a sufficient general form.

10.2   The Auxiliary Conformal Map

10.2.1   Definitions and Properties

Let Ω be a graph of a locally Lipschitz function $x_3 = f(x_1, x_2)$, defined over a domain $D \subset \mathbb{R}^2$, and let

$$ds_\Omega^2 = (1 + f_{x_1}'^2)\, dx_1^2 + 2 f_{x_1}'\, f_{x_2}'\, dx_1\, dx_2 + (1 + f_{x_2}'^2)\, dx_2^2 \tag{10.16}$$

be the square of its line element on Ω.

Fix arbitrarily a point $a \in D$ where f is differentiable. The function $f \in$ Lip D, and hence, by Rademacher's theorem, it is differentiable a.e. in D.

The quadratic form ds_Ω^2 is positive determined. Thus, an infinitesimal circle with respect to ds_Ω^2 and a center at a is an infinitesimal ellipse with respect to the Euclidean metric. Let $\theta(a)$, $0 \le \theta < \pi$, and $p(a)$, $p \ge 1$, be characteristics of this ellipse (i.e. the angle between its largest axis and the Ox_1-axis, and the ratio of the largest axis to the smaller, respectively).

It is not difficult to see that

$$\theta(a) = \frac{\pi}{2} + \operatorname{arctg} \frac{f_{x_2}'(a)}{f_{x_1}'(a)}, \quad p(a) = \sqrt{1 + |\nabla f(a)|^2}.$$

We consider a quasi-conformal map $u = u(x)$ of the domain D into the $u = (u_1, u_2)$-plane with characteristics which coincide a.e. in D with $\theta(x)$ and $p(x)$. Since $\operatorname{ess\,sup}_{D'} p(x) < \infty$ for every bounded subdomain D', $\overline{D'} \subset D$, then such map exists and defined up to conformal maps in the u-plane (see Theorem 3.49).

We assume that the following properties of quasi-conformal maps $u : D \to \mathbb{R}^2$ are known.

(1) A map $u = u(x)$ is differentiable a.e. in D, and at every point of the differentiability it transforms infinitesimal ellipses with characteristics $\theta(x)$, $p(x)$ onto infinitesimal circles (see [17, §1]).

(2) A map $u = u(x)$ has the first generalized Sobolev derivatives, square integrable locally in D, i.e. of $W_{\mathrm{loc}}^{1,2}(D)$ (see [17, §4]).

(3) Denote by $G = u(D)$ the image of D under $u = u(x)$, and by $x = x(u) = (x_1(u), x_2(u))$ - the inverse map. The characteristic $p(u)$ of the inverse map $x : G \to D$ is locally bounded in the domain G (see [17, §1]), and by the property (2) the map $x = x(u)$ belongs to $W_{\mathrm{loc}}^{1,2}(G)$ also.

We put $x_3(u) = f(x(u))$. Since f is locally Lipschitz in D, then by the property (3) the function $x_3(u) \in W_{\mathrm{loc}}^{1,2}(G)$.

From (2) it follows that the vector function

$$\chi(u) = (x_1(u), x_2(u), x_3(u))$$

also belongs to $W^{1,2}_{\mathrm{loc}}(G)$. It is clear that $\chi(u)$ is differentiable a.e.

The vector function $\chi : G \to \Omega$ realizes a bijective map of $G = u(D)$ onto Ω. Let $a \in G$ be a point where χ is differentiable. Denote by $j : F \to D$ the projection of Ω onto D. The map $\chi : G \to \Omega$ is a composition of maps j^{-1} and $x(u)$. Thus $\chi : G \to \Omega$ is conformal a.e. and a.e. on G it has properties (2.3). Variables u_1, u_2 are isothermal coordinates on Ω.

10.2.2 Conformal Types of Surfaces

By the stated above, we say that a surface Ω is of parabolic conformal type if $G = \mathbb{R}^2$ and of hyperbolic conformal type if $G \neq \mathbb{R}^2$.

Fix a function $h \in \mathcal{H}$. Denote by W the graph of the function

$$w(x) = \int\limits_{-\infty}^{f(x)} h(t)\,dt.$$

Theorem 10.17 *Let $D \subset \mathbb{R}^2$ be a simply connected domain bounded by a simple Jordan rectifiable curve. If f is a generalized solution of an equation (10.4) in D with structure restrictions (10.1), (10.2), (10.3), then for every $h \in \mathcal{H}$, the graph W of $x_3 = w(x)$ is of hyperbolic conformal type.*

Some indications of parabolicity and hyperbolicity of the conformal type of a function graph can be obtained from Theorem 1.36. Below we formulate two such indications.

Theorem 10.18 *Let f be a locally Lipschitz function and let Ω be its graph.*
If Ω is defined over a disc $|x| < R$, $0 < R \leq \infty$, and for a number $0 < r < R$ it is fulfilled

$$\iint\limits_{r < |x| < R} \frac{1 + \langle \nabla f, x^\perp \rangle^2}{\sqrt{1 + |\nabla f|^2}} \frac{dx_1 dx_2}{|x|^2} < \infty, \tag{10.19}$$

then Ω is of hyperbolic conformal type.
If Ω is defined over the plane $\mathbb{R}^2$ and

$$\lim_{R \to \infty} \frac{1}{R^2} \iint\limits_{1 < |x| < R} \frac{1 + \langle \nabla f, (x/|x|) \rangle^2}{\sqrt{1 + |\nabla f|^2}}\, dx_1 dx_2 = 0, \tag{10.20}$$

then Ω is of parabolic conformal type.

Here by $x^{\perp}$ we denote a unit vector in $\mathbb{R}^2$, orthogonal to the vector x and forming the angle $3\pi/2$ in the direction from $x^{\perp}$ to x.

The **proof** is not difficult. Fix $0 < r < R < \infty$. Let $\Gamma(r, R)$ be the family of all locally rectifiable arcs γ, joining boundary circles and lying in the ring $\{r < |x| < R\}$. Because $\operatorname{mod}_{\Omega} \Gamma(r, R)$ is the conformal invariant, then Ω is of parabolic type if and only if $\operatorname{mod}_{\Omega} \Gamma(r, R) = 0$. By the remark of Section 1.7, we have

$$\left(\int_r^R \frac{d\tau}{\tau \int_0^{2\pi} \frac{1 + \langle \nabla f, (x/|x|) \rangle^2}{\sqrt{1 + |\nabla f|^2}} \, d\theta} \right)^{-1} \geq$$

$$\geq \operatorname{mod}_{\Omega} \Gamma(r, R) \geq \int_0^{2\pi} \frac{d\theta}{\int_r^R \frac{1 + \langle \nabla f, x^{\perp} \rangle^2}{\tau \sqrt{1 + |\nabla f|^2}} \, d\tau}.$$

By the Cauchy integral inequality, we obtain

$$(R - r)^2 \leq \int_r^R \frac{d\tau}{\tau \int_0^{2\pi} \frac{1 + \langle \nabla f, (x/|x|) \rangle^2}{\sqrt{1 + |\nabla f|^2}} \, d\theta} \times$$

$$\times \int_r^R \tau \, d\tau \int_0^{2\pi} \frac{1 + \langle \nabla f, (x/|x|) \rangle^2}{\sqrt{1 + |\nabla f|^2}} \, d\theta.$$

Thus

$$\frac{(R - r)^2}{\iint\limits_{r < |x| < R} \frac{1 + \langle \nabla f, (x/|x|) \rangle^2}{\sqrt{1 + |\nabla f|^2}} \, dx_1 dx_2} \leq \int_r^R \frac{d\tau}{\tau \int_0^{2\pi} \frac{1 + \langle \nabla f, (x/|x|) \rangle^2}{\sqrt{1 + |\nabla f|^2}} \, d\theta}$$

that proves (10.20).

For the proof of (10.18), we observe that from Cauchy's inequality it follows

$$
\left(\int\limits_{0}^{2\pi} d\theta\right)^2 \leq \int\limits_{0}^{2\pi} \frac{d\theta}{\tau \int\limits_{r}^{R} \dfrac{1 + \langle \nabla f, x^{\perp}\rangle^2}{\sqrt{1 + |\nabla f|^2}}\, d\tau} \int\limits_{0}^{2\pi} \tau\, d\theta \int\limits_{r}^{R} \frac{1 + \langle \nabla f, x^{\perp}\rangle^2}{\sqrt{1 + |\nabla f|^2}}\, d\tau
$$

and

$$
\frac{4\pi^2}{\int\limits_{0}^{2\pi} d\theta \int\limits_{r}^{R} \dfrac{1 + \langle \nabla f, x^{\perp}\rangle^2}{\tau\,\sqrt{1 + |\nabla f|^2}}\, d\tau} \leq \int\limits_{0}^{2\pi} \frac{\tau\, d\theta}{\int\limits_{r}^{R} \dfrac{1 + \langle \nabla f, x^{\perp}\rangle^2}{\sqrt{1 + |\nabla f|^2}}\, d\tau}.
$$

Thus we deduce

$$
\frac{4\pi^2}{\iint\limits_{r<|x|<R} \dfrac{1 + \langle \nabla f, x^{\perp}\rangle^2}{\sqrt{1 + |\nabla f|^2}}\, \dfrac{dx_1 dx_2}{|x|^2}} \leq \int\limits_{0}^{2\pi} \frac{\tau\, d\theta}{\int\limits_{r}^{R} \dfrac{1 + \langle \nabla f, x^{\perp}\rangle^2}{\sqrt{1 + |\nabla f|^2}}\, d\tau}.
$$

The condition (10.19) implies that $\mathrm{mod}_\Omega \Gamma(r, R) > 0$ and Ω is of parabolic conformal type. $\qquad\square$

On indications of conformal type, see also [162], [126], [110], [51], [167] etc.

10.2.3 Some Properties of the Relative Distance

Let $\Omega \subset \mathbb{R}^3$ be a graph of a locally Lipschitz function f defined over a simply connected domain $D \subset \mathbb{R}^2$, and let $O' \in \Omega$ be a fixed point.

For a pair of points $p, q \in \Omega$, let $\rho(p, q; O', \Omega)$ be the relative distance, introduced in Chapter 4. By $\partial\tilde{\Omega}$, as above, we denote the boundary of Ω with respect to the metric ρ, i.e. it will be the set of all sequences $\{p_n\}$ of points in Ω, fundamental with respect to the relative metric ρ and having no accumulation points in Ω.

If Ω is the graph of the function $x_3 \equiv \mathrm{const}$ over the domain $D \subset \mathbb{R}^2$, then we identify Ω with D, and the relative boundary $\partial\tilde{\Omega}$ coincides with the Carathéodory boundary of D.

Suppose that D is simply connected and its image $G = u(D)$ is different from the entire plane $\mathbb{R}^2$. Because relations (2.3) are invariant under conformal transforms in the $u = (u_1, u_2)$-plane, then without loss of generality, we can assume that the domain G is bounded.

Consider the above described conformal map

$$\chi(u) = (x_1(u), x_2(u), x_3(u)) : G \to \Omega, \quad u = (u_1, u_2). \tag{10.21}$$

We put $O'' = \chi^{-1}(O')$ and denote by $r(G)$ the Euclidean distance from O'' to ∂G.

The following statement is a special case of Theorem 4.6.

Theorem 10.22   *If the area* $\mathrm{mes}_2(\Omega) < \infty$, *then for an arbitrary pair of points* $p, q \in G$, *satisfying the condition*

$$\rho(p, q; O'', G) < \min\{1, \frac{1}{16} r^4(G)\}, \tag{10.23}$$

it is fulfilled

$$\rho(\chi(p), \chi(q); O', \Omega) \leq K \log^{-1/2} \frac{1}{\rho(p, q; O'', G)}. \tag{10.24}$$

Here

$$K = 2\sqrt{\pi \, \mathrm{mes}_2(\Omega)}.$$

For the **proof**, it is sufficient to remark that the surface Ω, given by the graph of a locally Lipschitz function f, is locally bi-Lipschitz. It is obvious, since for every subdomain $D' \subset\subset D$, the relation (10.16) implies

$$|dx|^2 \leq ds_\Omega^2 = (1 + f_{x_1}'^2) \, dx_1^2 + 2f_{x_1}' f_{x_2}' \, dx_1 \, dx_2 + (1 + f_{x_2}'^2) \, dx_2^2 \leq C(f, D') \, |dx|^2$$

where $C(f, D')$ is a constant. $\square$

The estimate (10.24) implies that every fundamental (with respect to the relative distance $\rho(p, q; O'', G)$) sequence $\{a_n\} \subset G$ turns to a fundamental (with respect to the metric $\rho(\chi(p), \chi(q); O', \Omega)$) sequence $\{\chi(a_n)\} \subset \Omega$. Thus we obtain the following statement.

Corollary 10.25   *Under conditions described in Theorem 10.14, a conformal mapping (10.21) is continuously extended up to a continuous mapping of the relative boundary* $\partial \tilde{G}$ *onto the relative boundary* $\partial \tilde{\Omega}$.

It should be noted that the following important property of the projection $j : F \to D$.

Lemma 10.26 *Let Ω be the graph of a locally Lipschitz function, defined over a simply connected domain $D \subset \mathbb{R}^2$, and $j(O') = O$.*
  Then for every pair of points $p, q \in \Omega$, it fulfills

$$\rho(j(p),\, j(q); O, D) \leq \rho(p,\, q; O', \Omega)\,. \tag{10.27}$$

Proof. It is sufficient to remark that the projection $j : F \rightarrow D$ does not increase lengths of curves. $\qquad\qquad\square$

Let $x(u) = (x_1(u), x_2(u)) : G \rightarrow D$ be a map realized by components $x_1(u)$, $x_2(u)$ of the vector function (10.21).

Corollary 10.28 *If $\mathrm{mes}_2\,(\Omega) < \infty$, then for every pair of points $p, q \in G$ with the property (10.23), it fulfills*

$$\rho(x(p),\, x(q); O, D) \leq K \log^{-1/2} \frac{1}{\rho(p,\, q; O'', G)}$$

where K is the constant of Theorem 10.22.

The **proof** follows from Theorem 10.22 and Lemma 10.26. $\qquad\square$

10.3 Solution Jumps on a Boundary

We prove Theorem 10.10.
  At first we observe that the statement is trivial if $f \equiv \mathrm{const}$. Thus we can assume that $f \not\equiv \mathrm{const}$.

10.3.1 The Estimate of an Area Graph

Let f be a locally Lipschitz solution of the equation (10.4) with structure conditions (10.1)-(10.3). We assume that the domain D is simply connected and bounded with a simple Jordan rectifiable curve.

Lemma 10.29 *Under described suppositions, the following inequality holds*

$$\mathrm{mes}_2\,(W) \leq h_0 \nu + 2\,\mathrm{mes}_2\,(D)\,. \tag{10.30}$$

Proof. Fix a sequence of domains D_k, $k = 1, 2, \ldots$, such that

$$\overline{D}_k \subset D_{k+1}, \quad \cup_{k=1}^{\infty} D_k = D.$$

The function $w(x)$ belongs to the class Lip $(\overline{D}_k)$, $k = 1, 2, \ldots$. Choosing $\varphi = w(x)$ in (10.5), we have

$$\iint\limits_{D_k} \sum_{i=1}^{2} f'_{x_i} A_i(x, \nabla f)\, h(f(x))\, dx_1\, dx_2 = \int\limits_{\partial D_k} w(x) \sum_{i=1}^{2} A_i(x, \nabla f)\, n_i\, |dx|$$

$$\leq \int\limits_{\partial D_k} w(x)\, |A(x, \nabla f)|\, |dx|.$$

Since $0 < w(x) \leq h_0 < \infty$, then

$$\iint\limits_{D_k} \sum_{i=1}^{2} f'_{x_i} A_i(x, \nabla f)\, h(f(x))\, dx_1\, dx_2 \leq h_0 \int\limits_{\partial D_k} |A(x, \nabla f)|\, |dx|.$$

Passing to the limit as $k \to \infty$ and using the structure conditions (10.1)-(10.3), we find

$$\nu_1 \iint\limits_{D} \frac{|\nabla f|^2}{\sqrt{1 + |\nabla f|^2}}\, h(f(x))\, dx_1\, dx_2 \leq h_0\, \nu_2^*. \tag{10.31}$$

From (10.31) we deduce

$$\iint\limits_{D} \sqrt{1 + |\nabla f|^2} h(f(x))\, dx_1\, dx_2 \leq h_0 \frac{\nu_2^*}{\nu_1} + \iint\limits_{D} \frac{h(f(x))}{\sqrt{1 + |\nabla f|^2}}\, dx_1\, dx_2.$$

From here, remembering that $0 \leq h(f(x)) \leq 1$, we arrive to the estimate

$$\iint\limits_{D} |\nabla w(x)|\, dx_1\, dx_2 = \iint\limits_{D} |\nabla f(x)|\, h(f(x))\, dx_1\, dx_2 \leq h_0 \frac{\nu_2^*}{\nu_1} + \text{area}\,(D).$$

Because

$$\sqrt{1 + |\nabla w(x)|^2} < 1 + |\nabla w(x)|,$$

we find finally

$$\iint\limits_{D} \sqrt{1 + |\nabla w|^2}\, dx_1\, dx_2 \leq h_0 \frac{\nu_2^*}{\nu_1} + 2\,\text{area}\,(D)$$

that is equivalent to (10.30). $\qquad\qquad\qquad\qquad\qquad\qquad\qquad\qquad\square$

10.3.2 Monotonicity of Solutions

The square of the length element on the surface W is given by the formula

$$ds_W^2 = \sum_{i,j=1}^{2} g_{ij}(x)\, dx_i\, dx_j, \quad g_{ij}(x) = \delta_{ij} + w'_{x_i}\, w'_{x_j}\, dx_i\, dx_j \quad (i,j=1,2)$$

where δ_{ij} $(i,j=1,2)$ is the Kronecker symbol.

Let $(g^{ij}(x)) = (g_{ij})^{-1}(x)$ be the matrix inverse to (g_{ij}). Simple calculations imply

$$g^{ij} = \delta_{ij} - \frac{w'_{x_i} w'_{x_j}}{1 + |\nabla w|^2}.$$

For an arbitrary $\xi \in \mathbb{R}^2$, we put

$$|\xi|_W^2 = \sum_{i,j=1}^{2} g^{ij}(x)\, \xi_i \xi_j.$$

It is easy to check that

$$|\nabla w|_W^2 = \frac{1 + w'^2_{x_2}}{1 + |\nabla w|^2}\, w'^2_{x_1} - 2\frac{w'_{x_1} w'_{x_2}}{1 + |\nabla w|^2}\, w'_{x_1} w'_{x_2} +$$

$$+ \frac{1 + w'^2_{x_1}}{1 + |\nabla w|^2}\, w'^2_{x_2} = \frac{|\nabla w|^2}{1 + |\nabla w|^2}.$$

Since $\nabla w(x) = h(f(x))\,\nabla f(x)$, the relation (10.31) implies

$$\iint\limits_{D} \frac{|\nabla w|^2}{\sqrt{h^2(f) + |\nabla w|^2}}\, dx_1\, dx_2 =$$

$$\iint\limits_{D} \frac{|\nabla f|^2}{\sqrt{1 + |\nabla f|^2}}\, h(f(x))\, dx_1\, dx_2 \le h_0\, \frac{\nu_2^*}{\nu_1}.$$

From here, taking into account that $h(f(x)) \le 1$, we arrive to the inequality

$$\iint\limits_{D} \frac{|\nabla w|^2}{\sqrt{1 + |\nabla w|^2}}\, dx_1\, dx_2 \le h_0\, \frac{\nu_2^*}{\nu_1},$$

or

$$\iint\limits_{D} |\nabla w|_W^2 \sqrt{1 + |\nabla w|^2}\, dx_1\, dx_2 \le h_0\, \nu. \qquad (10.32)$$

Lemma 10.33 *Every generalized solution f of the equation (10.4) with structure conditions (10.1), (10.2), (10.3) is monotone in the Lebes- gue sense.*

Proof. Indeed, for example, suppose that $a \in D$ is a point of the strong
local maximum. For sufficiently small $\epsilon > 0$, the set $\{x \in D : f(x) > f(a) - \epsilon\}$
contains a precompact component $U \subset D$, $a \in U$.

We have $f(x) - f(a) + \epsilon = 0$ on the boundary of U. For a.e. $\epsilon > 0$, curves
∂U are rectifiable. Choosing in (10.5) the function $\phi = f(x) - f(a) + \epsilon$, for a.e.
$\epsilon > 0$ we can write

$$\iint\limits_{U} \sum_{i=1}^{2} f'_{x_i} A_i(x, \nabla f) \, dx_1 dx_2 =$$

$$\int\limits_{\partial U} (f(x) - f(a) + \epsilon) \sum_{i=1}^{2} A_i(x, \nabla f) \, n_i \, |dx| = 0.$$

Thus from (10.1), it follows

$$\iint\limits_{U} \frac{|\nabla f|^2}{\sqrt{1 + |\nabla f|^2}} \, dx_1 dx_2 = 0,$$

i.e. $\nabla f \equiv 0$ and $f \equiv \text{const}$ in U. We have the contradiction with the assumption
that a is the point of the strong local maximum. $\square$

10.3.3   Proof of Theorem 10.17

As above, with an auxiliary quasi-conformal map $u = u(x) : D \rightarrow \mathbb{R}^2$, we
introduce isothermal coordinates (u_1, u_2) on the surface W.

Consider the mapping $x = x(u) = (x_1(u), x_2(u))$, inverse to $u = u(x)$, and
the function $x_3(u) = w(x(u))$. The vector function

$$\chi(u) = (x_1(u), x_2(u), x_3(u))$$

realizes the one-to-one conformal mapping of $G = u(D)$ onto W.

Complete the proof of the theorem. At first we remark that by Lemma
10.33 the function $w(x)$ is monotone in D, and since the map $x(u) : D \rightarrow G$ is
homeomorphic, then $w^*(u) = w(x(u))$ is also monotone in G. Because $x(u)$ is
quasi-conformal and f is locally Lipschitz, then $w^* \in W^{1,2}_{\text{loc}}(G)$.

Suppose that the function w^* is defined over the whole of plane $\mathbb{R}^2$. Fix
$R > 0$ and $t > 1$. Let $B(\tau)$ and $S(\tau)$ be the disc and the circle with the center
at $u = 0$ and the radius τ, respectively.

As the proved above inequality (4.14), we prove that

$$\inf_{R\leq\tau\leq tR} \operatorname{osc}(S(\tau), w^*) \leq \left(\frac{2\pi\, I(R)}{\log t}\right)^{1/2} \tag{10.34}$$

where

$$I(R) = \iint\limits_{|u|<tR} |\nabla w^*|^2 du_1\, du_2.$$

The function w^* is monotone, and hence, for every $t > 1$, it fulfills

$$\operatorname{osc}(B(R), w^*) \leq \operatorname{osc}(S(R), w^*) \leq \inf_{R\leq\tau\leq tR} \operatorname{osc}(S(\tau), w^*).$$

Consequently,

$$\operatorname{osc}(B(R), w^*) \leq \left(\frac{2\pi\, I(\infty)}{\log t}\right)^{1/2}.$$

Taking into account (10.32), we find

$$I(\infty) = \iint\limits_{\mathbb{R}^2} |\nabla w^*|^2\, du_1\, du_2 = \iint\limits_{D} |\nabla w|_W^2 \sqrt{1 + |\nabla w|^2}\, dx_1\, dx_2 \leq h_0\, \nu,$$

and hence,

$$\operatorname{osc}(B(R), w^*) \leq \left(\frac{2\pi\, h_0\, \nu}{\log t}\right)^{1/2}.$$

Setting now $t \to \infty$, we conclude that $w^* \equiv \operatorname{const}$ on $B(R)$. But $R > 0$ is arbitrary, and hence, $w^* \equiv \operatorname{const}$ in the whole of plane $\mathbb{R}^2$. From here, it follows that $w \equiv \operatorname{const}$ in D.

However, if $w \equiv \operatorname{const}$ in D, then the map $u = u(x) : D \to \mathbb{R}^2$ is conformal with respect to the Euclidean metric, and the image of D can not be the whole of plane $\mathbb{R}^2$. We have a contradiction. $\qquad\square$

10.3.4 Points of Quasi-continuity

Because the simply connected domain $G \neq \mathbb{R}^2$ and the mapping $u(x) : D \to G$ is defined up to conformal mappings in the u-plane, we are right to assume that G is a unit disc with the center at O''.

We will show that *all points of ∂G are points of the quasi-continuity* of w^*. Fix a point $a \in \partial G$. Let $S(a, \tau)$ be a component of the set

$$\{u \in G : |u - a| = \tau\}$$

separating the point $a \in \partial G$ from the origin O'', $0 < \tau < 1$.

As in (10.34), we prove that for every $R \in (0, 1)$ and every $t \in (0, 1)$, there exists $\tau_0 \in (tR, R)$ such that

$$\inf_{tR < \tau < R} \operatorname{osc}\left(S(a, \tau), w^*\right) \leq \left(\frac{2\pi I}{\log \frac{1}{t}}\right)^{1/2}$$

where

$$I = \iint\limits_{G} |\nabla w^*|^2 \, du_1 \, du_2 \,.$$

Choosing now $t_n \to 0$, we find a sequence $\tau_n \to 0$ along which

$$\operatorname{osc}(S(a, \tau_n), w^*) \to 0 \,. \tag{10.35}$$

This means that a is the quasi-continuity point of w^* at $a \in \partial G$.

By Lemma (10.29) the integral

$$\iint\limits_{D} \sqrt{1 + |\nabla w|^2} \, dx_1 \, dx_2 < \infty \,.$$

Using Theorem 10.14 and Lemma 10.26, we conclude that the mapping

$$x(u) = j \circ \chi(u) : G \to D$$

is continuous up to the boundary of the unit disc ∂G.

Denote by $\tilde{x}(u) : \overline{G} \to \overline{D}$, $\tilde{x}(u)|_G = x(u)$, the map between closed domains obtained with the continuous extension of $x(u) : G \to D$ to ∂G.

Moreover, by Corollary 10.28 for arbitrary points

$$p, q \in \overline{G} \quad \text{with} \quad \rho(p, q; O'', G) < 1/16 \,,$$

we have

$$\rho(\tilde{x}(p), \tilde{x}(q); O, D) \leq K_1 \log^{-1/2} \frac{1}{\rho(p, q; O'', G)}$$

where

$$K_1 = 2\sqrt{\pi \operatorname{mes}_2 (W)}$$

is the constant.

By Lemma 10.29

$$K_1 \leq 2\sqrt{\pi} \left(\nu \, h_0 + 2 \operatorname{area}(D)\right)^{1/2} = K \,.$$

In particular, if points p and q lie on the boundary circle and $|p - q| < 1/16$, then $\rho(p, q; O'', G) = |p - q|$. Thus the found estimate takes the form

$$\rho(\tilde{x}(p), \tilde{x}(q); O, D) \leq K \log^{-1/2} \frac{1}{|p - q|} \qquad |p - q| < \frac{1}{16} \,. \tag{10.36}$$

Since the mapping $\tilde{x}(u) : \overline{G} \to \overline{D}$ is continuous, then the preimage $\tilde{x}^{-1}(a)$ of a point $a \in \partial D$ is a closed set on the unit circle ∂G. Moreover, for every pair of points $a \neq b$ on ∂D, it is fulfilled

$$\tilde{x}^{-1}(a) \cap \tilde{x}^{-1}(b) = \emptyset .$$

The set $\tilde{x}^{-1}(a)$ is connected. Indeed, because ∂D is the simple Jordan curve, for sufficiently small $\epsilon > 0$ the set $U_\epsilon = \{x \in D : |x - a| < \epsilon\}$ is connected. The mapping $x(u) : G \to D$ is homeomorphic, and hence, preimages $x^{-1}(U_\epsilon)$ are connected for all $\epsilon \in (0, \epsilon_0)$ where $\epsilon_0 > 0$ is a sufficiently small number. Thus the set

$$\tilde{x}^{-1}(a) = \cap_{\epsilon > 0} \overline{U}_\epsilon$$

is also connected.

So, for an arbitrary point $a \in \partial D$, the set $\tilde{x}^{-1}(a)$ can be either an isolated point or a closed connected arc.

Since the amount of non-overlapping arcs on a circle is no more than countable, then the inverse mapping $\tilde{x}^{-1}(x)$ is defined everywhere on ∂D, except, at most, in a countable set $E \subset \partial G$.

We show that *every point $a \in \partial D \setminus E$ is the quasi-continuity point* of the function $w(x)$.

Let $b = \tilde{x}^{-1}(a)$ be the preimage of the point $a \in \partial D \setminus E$. By (10.35) there exists a sequence of subdomains $\{G_n\}$, $\partial' G_n = S(b, \tau_n)$ with properties:

every arc $\partial' G_n$ separates b from the origin O;

length $(\partial' G_n) \to 0$ and osc $(\partial' G_n, w^*) \to 0$ as $n \to \infty$;

finally, $\cap_{n=1}^\infty [G_n] = \emptyset$.

We put $D_n = x(G_n)$, $n = 1, 2, \ldots$. Because the map $x(u) : G \to D$ is homeomorphic, every arc $\partial D_n = x(S(b, \tau_n))$ separates $a \in \partial D$ from O.

The map $\tilde{x}(u) : \overline{G} \to \overline{D}$ is continuous, and hence, length $\partial' D_n \to 0$ and $\cap_{n=1}^\infty [D_n] = \emptyset$.

Finally, osc $(\partial' D_n, w) = $ osc$(S(b, \tau_n), w^*)$, and hence, osc $(\partial' D_n, w) \to 0$ as $n \to \infty$.

Thus the point $a \in \partial D \setminus E$ has all properties, necessary to define the points of quasi-continuity of $w(x)$.

10.3.5 Behavior of Solutions at Jump Points

Fix a jump point $a \in E$. Its preimage $\tilde{x}^{-1}(a)$ is a subarc β of the circle $|u| = 1$. We estimate its length $l(\beta)$. Let $\xi \in \beta$ be the middle of the arc. It is easy now to calculate

$$r = 2\sin\frac{l(\beta)}{4},$$

which would be the distance from ξ to arc ends.

Suppose that $r < 1$, i.e. $l(\beta) < \frac{2}{3}\pi$. Consider the family of circles $\{S(\xi, \tau)\}$ with the center at ξ and radii τ, $r < \tau < 1$. We put $C_\tau = G \cap S(\xi, \tau)$.

By (4.12) we have

$$\int_r^1 \frac{l^2(\chi(C_\tau))}{\tau}\, d\tau \leq 2\pi \iint_{\Delta_{r,1}} \sum_{i=1}^3 |\nabla x_i(u)|^2\, du_1\, du_2.$$

Remark now that for every $\tau \in [r, 1]$, it is fulfilled

$$l(\chi(C_\tau)) \geq \max\{\operatorname{osc}(x(C_\tau), w), \operatorname{length}(x(C_\tau))\} \geq \delta(a, w, O),$$

and hence,

$$\delta^2(a, w, O) \leq 4\pi \log^{-1}\frac{1}{r}\operatorname{mes}_2(W).$$

Thus, taking into account the estimate (10.30), we arrive to the inequality

$$\delta^2(a, w, O) \leq \frac{K^2}{\log 1/r}$$

where K is the constant of Theorem 10.10.

From here,

$$r = 2\sin\frac{l(\beta)}{4} \geq \exp\{-\frac{K^2}{\delta^2(a, w, O)}\},$$

and we have the following estimate of $l(\beta)$ for 'small' $\delta(a, w, O)$:

$$l(\beta) \geq 4\arcsin\left(\frac{1}{2}\exp\{-\frac{K^2}{\delta^2(a, w, O)}\}\right) \quad \text{for} \quad l(\beta) < \frac{2}{3}\pi. \tag{10.37}$$

Find the estimate of $l(\beta)$ for 'big' $\delta(a, w, O)$.

Let $a \in \partial D \cap E$ be a jump point in which $\frac{2}{3}\pi \leq l(\beta) \leq 2\pi$. Without loss of generality, we can assume that the arc $\beta = \tilde{x}^{-1}(a)$ is described by relations

$$\beta = \{u = (u_1, u_2) : u_1^2 + u_2^2 = 1, -\frac{l(\beta)}{2} \leq \operatorname{arctg}\frac{u_2}{u_1} \leq \frac{l(\beta)}{2}\}.$$

Fix the segment

$$\gamma = \{u = (u_1, u_2) : 0 \leq u_1 \leq 1, u_2 = 0\}$$

and denote by p_1 and p_2 the boundary points of the unit disc with the cut γ, lying in the intersection of the circle $|u| = 1$ and the upper and lower edges of γ respectively.

Let $T : G \to V$ be the conformal mapping of the disc $|u| < 1$ with the cut γ onto the half-disc $V = \{v = (v_1, v_2) : v_1^2 + v_2^2 < 1, v_2 > 0\}$ for which $T(p_1) = (1, 0)$, $T(p_2) = (-1, 0)$ and $T(0,0) = (0,0)$.

Under the map T, the arc $\beta \subset \partial G$ is corresponding to

$$\eta = \left\{ v = (v_1, v_2) : \frac{l(\beta)}{4} \leq \operatorname{arctg} \frac{v_2}{v_1} \leq \pi - \frac{l(\beta)}{4} \right\}.$$

For every $0 < k < \cos \frac{l(\beta)}{4}$, the rectilinear segment

$$\zeta(k) = \{v = (v_1, v_2) : -\sqrt{1 - k^2} < v_1 < \sqrt{1 - k^2}, \ v_2 = k\}$$

separates in the domain V the arc η from the point $(0,0)$. Its image $\zeta^*(k) = \tilde{x} \circ T^{-1}(\zeta(k))$ is an arc, separating points a and O in D. Thus for every $k \in (0, \cos \frac{l(\beta)}{4})$ it is fulfilled

$$l^2(\zeta^*(k)) \leq \left(\int_{\zeta(k)} \left| \frac{\partial \chi^*}{\partial v_1} \right| dv_1 \right)^2 \leq 2\sqrt{1 - k^2} \int_{\zeta(k)} \left| \frac{\partial \chi^*}{\partial v_1} \right|^2 dv_1$$

where $\chi^* = \chi \circ T^{-1}$.

Further,

$$\int_0^{\cos l(\beta)/4} \frac{l^2(\zeta^*(k))}{\sqrt{1 - k^2}} dk \leq 2 \iint_V \left| \frac{\partial \chi^*}{\partial v_1} \right|^2 dv_1 \, dv_2 \leq 2 \operatorname{mes}_2 (W).$$

Taking into account that

$$l(\zeta^*(k)) \geq \delta(a, w, O) \quad \text{for all} \quad k \in (0, \cos \frac{l(\beta)}{4}).$$

From here we obtain

$$\delta^2(a, w, O) \int_0^{\cos l(\beta)/4} \frac{1}{\sqrt{1 - k^2}} dk \leq 2 \operatorname{mes}_2 (W),$$

or

$$\delta^2(a, w, O) \left(\frac{\pi}{2} - \frac{l(\beta)}{4} \right) \leq \frac{K^2}{2\pi}$$

where K is the constant of Theorem 10.10.

Thus we arrive to the estimate

$$l(\beta) \geq \max\left\{2\pi - \frac{2K^2}{\pi\,\delta^2(a, w, O)}, 0\right\}. \tag{10.38}$$

10.3.6   Estimate of a Summary Jump

Let a_k be jump points of the function $w(x)$ and let β_k be corresponding closed arcs on $|u| = 1$, $k = 1, 2, \ldots$. We are right to assume that

$$l(\beta_1) \geq l(\beta_2) \geq \ldots \geq l(\beta_k) \geq \ldots.$$

Since $\beta_i \cap \beta_j = \emptyset$ for $i \neq j$, then

$$\sum_{k=1}^{\infty} l(\beta_k) \leq 2\pi.$$

Beginning at least with $k = 4$, it is fulfilled $l(\beta_k) < \frac{2}{3}\pi$, and we can use the estimate (10.37). We have

$$4\sum_{k=4}^{\infty} \arcsin\left(\frac{1}{2}\exp\{-\frac{K^2}{\delta^2(a, w, O)}\}\right) \leq \sum_{k=4}^{\infty} l(\beta_k). \tag{10.39}$$

For every point a_1, a_2, a_3, we use the estimate (10.38). Then

$$\sum_{k=1}^{3} \max\{2\pi - \frac{2K^2}{\pi\delta^2(a, w, O)}, 0\} \leq \sum_{k=1}^{3} l(\beta_k). \tag{10.40}$$

Combining (10.39) and (10.40), we arrive to the inequality

$$\sum_{k=1}^{3} \max\{2\pi - \frac{2K^2}{\pi\,\delta^2(a, w, O)}, 0\} + 4\sum_{k=4}^{\infty} \arcsin\left(\frac{1}{2}\exp\{-\frac{K^2}{\delta^2(a, w, O)}\}\right)$$

$$\leq \sum_{k=1}^{3} l(\beta_k) + \sum_{k=4}^{\infty} l(\beta_k) \leq 2\pi.$$

Thus Theorem 10.10 is proved completely.   □

10.4 The Fatou Type Theorem

Prove Theorem 10.7. Let $f : D \to \mathbb{R}$ be a function, monotone in the Lebesgue sense, and let $a \in \partial D$ be a point. We call that f *has the Lindelöf property at* a if the existence finite limits of f along ways Γ_1 and Γ_2, leading to this point, implies the existence of a limit of f along every way, leading to a and lying between Γ_1 and Γ_2.

The proof of Theorem 10.7 is based on the following statement.

Lemma 10.41 *If a function is monotone in the Lebesgue sense and quasi-continuous at a boundary point, then this function has the Lindelöf property at this point.*

For the **proof** we fix a point of quasi-continuity $a \in \partial D$ and a sequence of subdomains $\{D_k\}_{k=1}^{\infty}$ of D with properties (α), (β), (γ) of the definition 10.9.

Let Γ_1 and Γ_2 be arbitrary ways, leading to a, along which f has limits α_1 and α_2 respectively. Denote by U_k the subdomain of D_k lying between Γ_1, Γ_2, ∂D_k and ∂D_{k+1}. Without loss of generality, we can assume that such subdomain is defined for every $k = 1, 2, \dots$.

For $k = 1, 2, \dots$, we put

$$\gamma_{1k} = \partial U_k \cap \Gamma_1, \quad \gamma_{2k} = \partial U_k \cap \Gamma_2, \quad \gamma_{3k} = \partial U_k \cap \partial D_k, \quad \gamma_{4k} = \partial U_k \cap \partial D_{k+1}.$$

The function f has finite limits along Γ_1 and Γ_2. Therefore,

$$\mathrm{osc}\,(\gamma_{ik}, f) \to 0 \quad \text{as} \quad k \to \infty \quad \text{for} \quad i = 1, 2.$$

Moreover, by (γ) of the definition 10.9, we have

$$\mathrm{osc}\,(\gamma_{ik}, f) \to 0 \quad \text{as} \quad k \to \infty \quad \text{for} \quad i = 3, 4.$$

Thus, taking into account that

$$\partial U_k = \cup_{i=1}^{4} \gamma_{ik},$$

we can conclude

$$\mathrm{osc}\,(\partial U_k, f) \to 0 \quad \text{as} \quad k \to \infty.$$

However, f is monotone in the Lebesgue sense, and therefore,

$$\mathrm{osc}\,(U_k, f) \to 0 \quad \text{as} \quad k \to \infty.$$

This means that f has a limit along the sequence $\{U_k\}$ and, in particular, $\alpha_1 = \alpha_2$. Thus f has the Lindelöf property at a. $\qquad \square$

Proof of Theorem 10.7. Let f be a solution of (10.4) with structure restrictions (10.1), (10.2), and (10.3) in D. We put $\Phi(x) = \operatorname{arctg} f(x)$.

Choose $h(t)$, as in Example 10.13. By (10.30), we have

$$\iint\limits_{D} |\nabla\Phi(x)|\, dx_1\, dx_2 < \infty.$$

Denote by D_c the intersection of D and the line $\{(x_1, x_2) \in \mathbb{R}^2 : x_1 = c\}$, and by Γ_c - the intersection of ∂D and the same line. Using Fubini theorem, we conclude that for a.e. c, belonging to the projection of D onto the axis $0x_1$, the following relation holds

$$\int\limits_{D_c} |\nabla\Phi(c, x_2)|\, dx_2 < \infty.$$

From here for every pair of points (c, x_2'), (c, x_2'') which belong to one and the same connected component of D_c, we obtain

$$|\Phi(c, x_2'') - \Phi(c, x_2')| \leq \int\limits_{x_2'}^{x_2''} |\nabla\Phi(c, x_2)|\, dx_2 \quad (x_2' < x_2'').$$

Thus, $\Phi(c, x_2)$ is uniformly continuous on every component of the set D_c, and hence, at every point $a \in \Gamma_c$, it has a limit along D_c.

We turn the coordinate system in the (x_1, x_2)-plane in the angle $\alpha \in F$ where F is a countable set dense everywhere on $(0, 2\pi)$, and reason every time as above. Thus we can touch a.e. point $a \in \partial D$ with a corner of any angle, close to π and such that Φ has limits along every side of this corner.

Suppose that $a \in \partial D$ be a point of the described form. By Theorem 10.10 the function $\Phi = \operatorname{arctg} f$ is quasi-continuous everywhere on ∂D except, possibly, in a countable set, and without loss of generality, we are right to assume that a is a point of the quasi-continuity of f. By Lemma 10.33 the function f (and, hence, $\Phi = \operatorname{arctg} f$) is monotone in the Lebesgue sense. Thus by Lemma 10.41, this function has Lindelöf's property at a.

Choose arbitrarily a corner along sides of which Φ has limits. These limits equal a number α, $|\alpha| \leq \frac{\pi}{2}$, and the constant α is also the limit of Φ along the corner domain. But magnitude of this corner can be chosen arbitrarily close to π, and hence, Φ has the angular boundary value α.

From here, $f = \operatorname{tg}\Phi$ has the finite or infinite angular boundary value $\operatorname{tg}\alpha$ at this point. $\square$

10.5   Continuity and Quasi-continuity

Below we prove Theorem 10.22. We will need a special case of Finn's lemma [38]. In the form, necessary below, this statement is contained in [141, Lemma **10.2**].

Lemma 10.42   *Let Δ be a subdomain, lying in a ring $1 < |x| < b$, and let γ be a set of boundary points of Δ which are not lying on the unit circle. Let f be a solution of the minimal surface equation (2.20) in Δ such that, for all points $x \in \gamma$*

$$m \le f(x) \le M$$

where m and M are constants.
Then, everywhere in Δ, the following inequality holds

$$m - \operatorname{arccosh} b \le f(x) \le M + \operatorname{arccosh} b. \tag{10.43}$$

Proof of Theorem 10.14. The function f can not be identically to $\pm\infty$ on an arc $\Gamma_1 \subset \Gamma$ (see [141, §10]). Thus there exists an everywhere dense set on Γ, such that in its every point, it is possible to touch outwards D with an arc along which f is bounded. Using Lemma 10.42, from here we conclude that in arbitrary strongly inner subarc $\Gamma_1 \subset \Gamma$, limiting values of f are bounded. Thus it is sufficient to prove the statement in case if D is simply connected and bounded by a simple Jordan rectifiable curve and f is bounded in D.

By Theorem 10.10 the function f is quasi-continuous everywhere on ∂D, except, possibly, in a countable set.

Fix a point $a \in \partial D$ where f is quasi-continuous, and an inner point $O \in D$. There exists a sequence $\{\gamma_k\}$, $\gamma_k \subset D$, $k = 1, 2, \ldots$, of arcs with ends on Γ separating the point a from O and such that

$$\lim_{k \to \infty} \operatorname{osc}(\gamma_k, f) = 0. \tag{10.44}$$

Denote by D_k the subset of D separating γ_k from the point O. The inequality (10.43) implies that

$$\operatorname{osc}(D_k, f) \le \operatorname{osc}(\gamma_k, f) + 2\operatorname{arccosh}(1 + \operatorname{length}(\gamma_k)) \quad (k = 1, 2, \ldots).$$

Thus by (10.44) we obtain

$$\lim_{k \to \infty} \operatorname{osc}(D_k, f) = 0,$$

and f is continuously extended to the point $a \in \Gamma$. Theorem is proved. $\qquad \square$

Open questions 10.45   1) Find a direct (not using auxiliary conformal mappings of a surface to the plane of isothermal coordinates) proof of Theorem 10.10.   2) Study behavior of generalized solutions near points of quasi-continuity and near jump points. In what cases, do limits of solutions along non-tangent ways exist? Along tangent ways?   3) Find multi-dimensional versions of Fatou's type theorem for minimal graphs.

Index

251

References

[1] D.R. Adams and L.I. Hedberg, *Function Spaces and Potential Theory*, Springer-Verlag, Berlin-Heidelberg-New York etc., 1996.

[2] L.V. Ahlfors, *Untersuchungen zur Theorie der konformen Abbildungen und der ganzen Funktionen*, Acta. Soc. Sci. Fenn. N. S. A I, **1**, n. 9, 1930, 1-40.

[3] L.V. Ahlfors, *Lectures on quasiconformal mappings*, D. Van Nostrand Comp. Inc., Princeton, 1966.

[4] L.V. Ahlfors, *Conformal invariants*, McGraw-Hill, New York, 1973.

[5] R.S. Akopyan, *Stabilization conditions of a minimal surface over half-strip*, in "Scien. schools of Volgograd State univ. Geometric analysis and its applications", izd-vo VolGU, Volgograd, 1999, 105-110 (in Russian).

[6] G. Alessandrini and V. Nessi, *Univalent σ-harmonic mappings*, Arch. Ration. Mech. and Anal., v. 158, 2001, 155-171.

[7] G.S. Asanov, *Finslerlike Geometry*, izd-vo MGU, Moscow, 2004 (in Russian).

[8] K. Astala, *Area distortion of quasiconformal mappings*, Acta Math., **173**, 1994, 37—60.

[9] K. Astala and V. Nesi, *Composites and quasiconformal mappings: new optimal boundes*, Preprint 223, Departm. of Math., University of Jyväskylä, 2000, 26 pp.

[10] F.G. Avkhadiev, *Conformal maps and boundary problems*, Kazan' fund "Mathematics", 1996, 216 pp.

[11] C. Bandle, *Isoperimetric inequalities and applications*, Pitman Adv. Pub. Program: Boston—London—Melbourne, 1980.

[12] E.F. Beckenbach, R. Bellman, *Inequalities*, Springer-Verlag, Berlin-Göttingen-Heidelberg, 1961.

[13] P.P. Belinskii, *On distortion under quasiconformal maps*, Dokl. AN SSSR, v. 91, n. 5, 1953, 997-998 (in Russian).

[14] P.P. Belinskii, *On area measure under quasiconformal mappings*, Dokl. AN SSSR, v. 121, n. 1, 1958, 16-17 (in Russian).

[15] P.P. Belinskii, *Solution of extremal problems of the quasiconformal maps theory by the variational method*, Sib. math. j., v. 1, n. 3, 1960, 303-330 (in Russian).

[16] P.P. Belinskii, *On the closeness order of a space quasiconformal map to the conformal map*, Sib. math. j., v. 14, n. 3, 1973, 475-483 (in Russian).

[17] P.P. Belinskii, *General properties of quasiconformal maps*, Nauka, Sib. otd., Novosibirsk, 1974 (in Russian).

[18] V.I. Belyi, V.M. Miklyukov, *Some properties of conformal and quasiconformal maps and direct theorems of the constructive function theory*, Izv. AN SSSR, Ser. math., v. 38, n. 6, 1974, 1343-1361 (in Russian).

[19] S.N. Bernstein, *On a geometrical theorem and its applications to partial differential equations*, Sobr. soch., v. III, Izd-vo AN SSSR, 1960, 251-258 (in Russian).

[20] L. Bers, *Mathematical aspects of subsinic and transonic gas dinamics*, New York · John Wiley & Sons, inc. London · Chapman & hall, Limited, 1958.

[21] L. Bers, *Uniformization by Beltrami equation*, Comm. Pure Appl. Math., v. 14, 1961, 215-228.

[22] A.A. Borisenko, *The inner and exterior geometry of multi dimensional submanifolds*, Izd-vo Ekzamen, Moscow, 2003 (in Russian).

[23] C. Caratheodory, *Untersuchungen über die konformen Abbildungen von festen und veränderlichen Gebieten*, Math. Ann., v. 72, 1912, 107-144.

[24] C. Caratheodory, *Über die Begrenzung einfach zusammenhangender Gebiete*, Math. Ann., v. 73, 1913, 323-370.

[25] C. Caratheodory, *Conformal Representation*, Cambridge at the University Press, 1932.

[26] T. Carleman, *Sur une inégalité differentielle dans la théorie des functions analytiques*, C.R.Acad. Sci. Paris, v. 196, 1933, 995-997.

[27] S. Chandrasekhar, *The mathematical theory of black holes*, Clarendon Press Oxford Oxford University Press New York, 1983.

[28] E.F. Collingwood, A.J. Lohwater, *The theory of cluster sets*, Cambrige Univ. Press, Cambrige, 1966.

[29] R. Courant, *Dirichlet's principle, conformal mapping, and minimal surfaces*, Interscience, New York, 1950.

[30] G. David and P. Mattila, *Removable sets for Lipschitz harmonic functions in the plane*, Revista Matem. Iberoamericana, v. 16, n. 1, 2000, 137-215.

[31] E.P. Dolgenko, *On singularity removal of analytic functions*, Uspehi math. nauk, v. 18, n. 4, 1963, 135-142 (in Russian).

[32] Yu.V. Dymchenko, *Equality of capacity and module of a condenser on a surface*, Zapiski nauchn. seminarov LOMI, v. 276, 2001, 112-133 (in Russian).

[33] J. Eells and B. Fuglede, *Harmonic Maps between Riemannian Polyhedra*, Cambridge Tracts in Mathematics **142**, Cambridge Univ. Press, UK, 2001.

[34] A. Eremenko and D.H. Hamilton, *On the area distortion by quasiconformal mappings*, Proceedings of the American Mathematical Sosiety, **123**, n. 9, 1995, 2793-2797.

[35] M.A. Evgrafov, *Analytic functions*, Nauka, Moscow, 1968 (in Russian).

[36] D. Faraco, *Beltrami operators and microstructure*, Academic dissertation, Depart. of Math., Faculty of Sci., University of Helsinki, Helsinki, 2002.

[37] H. Federer, *Geometric Measure Theory*, Springer-Verlag, Berlin, 1969.

[38] R. Finn, *Remarks relevant to minimal surfaces, and to surfaces of prescribed mean curvature*, J. Analyze Math., v. 14, 1965, 139-160.

[39] A.T. Fomenko, *Variational methods in the topology*, Nauka, Moscow, 1982 (in Russian).

[40] B. Fuglede, *Extremal length and functional completetion*, Acta Math., v. 98, n.3-4, 1957, 171-219.

[41] F.R. Gantmacher, *Theory of matrices*, Nauka, Moscow, 1967 (in Russian).

[42] F.W. Gehring, O. Lehto, *On the total differentiability of functions of a complex variable*, Ann. Acad. Sci. Fenn. A I, **272**, 1959, 1-9.

[43] F. Gehring and E. Reich, *Area distortion under quasiconformal mappings*, Ann. Acad. Sci. Fenn. Ser. AI **388**, 1966, 1-14.

[44] E. Giusti, *Minimal surfaces and functions of bounded variation*, Birkhäuser, Boston-Basel-Stuttgart, 1984.

[45] S.K. Godunov, E.I. Romenskii, *Elements of mechanics and conservation laws*, Izd-vo "Nauchnaya kniga", Novosibirsk, 1998 (in Russian).

[46] V.M. Goldstein, Yu.G. Reshetnyak, *Introduction to theory of functions with generalized derivatives and quasiconformal maps*, Nauka, Moscow, 1983 (in Russian).

[47] G.M. Golusin, *Geometric theory of complex variable functions*, Nauka, Moscow, 1966 (in Russian).

[48] C. Grammatico, *A result on strong unique continuation for the Laplace operator*, Commun. in partial defferential equations, v. 22 (9&10), 1997, 1475-1491.

[49] W. Gresky, *Konforme Abbildungen der Oberfläche Eines Rektangulären Hexaeders auf die Kugeloberfläche*, Inaugural-Dissertatin zur Erlangung def Doktorwürde der Hohen Philosophischen Fakultät der Universität Leipzig, Weida i. Thür, 1928, 1-74.

[50] E.G. Grigoryeva, A.A. Klyachin, and V.M. Miklyukov, *Problem of Functional Extension and Space-Like Surfaces in Minkowski Space*, Journals of Analysis and its Applications, v. 21, n. 3, 2002, 719-752.

[51] A. Grigor'yan, *Analytic and geometric background of recurrence and non-explosion of the Brownian motion on Riemannian manifolds*, Bull. Amer. Math. Soc., v. 36, n. 2, 1999, 135-249.

[52] M. Gromov, *Metric Structures for Riemannian and Non-Riemannian Spaces*, Boston-Basel-Berlin, Birkhäuser, 1999.

[53] I.M. Grudskii, *Structure of inner coordinates on combined Riemannian surfaces*, In "Differential, integral equations and complex analysis", Elista, 1986, 30-45 (in Russian).

[54] I.M. Grudskii, *Christoffel-Schwarz formula for polyhedral surfaces*, Dokl. AN SSSR, v. 307, n. 1, 1989, 15-17 (in Russian).

[55] V. Gutlyanskii, O. Martio, T. Sugawa and M. Vuorinen, *On the Degenerate Beltrami Equation*, Reports of the Department of Mathematics, University of Helsinki, Preprint 282, 2001, 32 pp.

[56] P. Hajlasz, *Sobolev Mappings, Co-Area Formula and Related Topics*, Proceedings on analysis and geometry, Sobolev Institute Press, Novosibirsk, 2000, 227-254.

[57] H. Haken, *Advanced Synergetics*, Springer - Verlag, Berlin Heidelberg New York Tokyo, 1983.

[58] S.Ya. Havinson, *On erasure of singularities*, Litovskii math. sb., III, n. 1, 1963, 271-287 (in Russian).

[59] J. Heinonen, T. Kilpeläinen, and O. Martio, *Nonlinear potential theory of degenerate elliptic equations*, Clarendon Press, Oxford etc., 1993.

[60] J. Heinonen, P. Koskela, *Sobolev Mappings with Integrable Dilatations*, Arch. Rational Mach. Anal., v. 125, 1993, 81-97.

[61] I.O. Herzog, *Phragmén-Lidelöf theorems for second order quasilinear elliptic partial differential equations*, Proc. Amer. Math. Soc., v. 15, 1964, 721-728.

[62] D.P. Il'yutko, *Locally minimal nets in N-normed spaces*, Math. zametki, v. 74, n. 5, 2003, 656-668 (in Russian).

[63] O.Yu. Imanuvilov, *On Carleman estimates for hyperbolic equations*, Asymptotic Analysis, v. 32, 2002, 185-220.

[64] T. Iwaniec, G. Martin, *Geometric Function Theory and Nonlinear Analysis*, Oxford Mathematical Monographs, Oxford University Press, Oxford, 2001.

[65] O.V. Ivanov, G.D. Suvorov, *Full lattice of conformally invariant compactifications of a domain*, Naukova Dumka, Kiew, 1982 (in Russian).

[66] J.A. Jenkins, *Univalent functions and conformal mapping* , Springer-Verlag, Berlin-Göttingen-Heidelberg, 1958.

[67] A.P. Karmazin, *Quasiisometries, theory of preends and metric structures of space domains*, Izd-vo Surgutskogo un-ta, Surgut, 2003 (in Russian).

[68] R. Kersner, A. Shishkov, *Instantaneous Shrinking of the Support of Energy Solutions*, J. of Math. Anal. and Appl., v. 198, 1996, 729-750.

[69] V.A. Klyachin, V.M. Miklyukov, *Tubes and bands in the space-time*, Izd-vo Volgogradsk. univ., Volgograd, 2004 (in Russian).

[70] A.P. Kopylov, *Stability of Classes of Mappings and Hölder Continuity of Higher Derivatives of Elliptic Solutions to Systems of Nonlinear Differential Equations*, Sib. Math. J., v. 43, n. 1, 2002, 68-82 (in Russian).

[71] P. Koskela, J. Malý, *Mappings of finite distortion: The zero set of the Jacobian*, Journal of the European Mathematical Society, v. 5, n. 2, 2003, 95-105.

[72] V.I. Kruglikov, *On existence and uniqueness of maps quasiconformal in mean*, in 'Metr. voprosy theor. func. i otobr.', n. IV, Naukova Dumka, Kiew, 1973, 123-147 (in Russian).

[73] V.I. Kruglikov, *Condenser capacities and space maps quasiconformal in mean*, Math. sb., v. 130 (172), n. 2, 1986, 185-206 (in Russian).

[74] V.I. Kruglikov, V.M. Miklyukov, *Stability theorems for maps of the class BL*, in 'Metr. voprosy theor. functii i otobr.', n. III, Izd-vo Naukova Dumka, Kiew, 1971, 55-70 (in Russian).

[75] V.I. Kruglikov, V.M. Miklyukov, *On some classes of plane topological maps with generalized derivatives*, in 'Metr. voprosy theor. functii i otobr.', n. IV, Izd-vo Naukova Dumka, Kiew, 1973, 102-122 (in Russian).

[76] S.L. Krushkal', *On mappings quasiconformal in mean*, Sib. math. j., v. 8, n. 4, 1967, 798-806 (in Russian).

[77] S.L. Krushkal', *Quasiconformal mappings and Riemann surfaces*, John Wiley&Sons, New York-Toronto-London-Sydney, 1979.

[78] N. Kuiper, *On C^1-isometric embeddings*, 1, 2, Proc. Koninkl. Nederl. Akad. Wetensch. Ser. A, v. 58, n. 4, 5, 1955, 545-556, 683-689.

[79] B.P. Kufarev, *Potentials and boundary correspondence*, Izv. AN SSSR, Ser. math., v. 41, n. 2, 1977, 438-461 (in Russian).

[80] R. Kühnau, *Trianguliere Riemannsche Mannigfaltigkeiten mit ganz-linearen Bezugssub-Stitutionen and st"uchweise konstanter komplexer Dilatation*, Math. Nachrichten, B. 46, Heft 1-6, 1970, 243-261.

[81] R. Kühnau, *Über zwei klassen schlichter konformer Abbildungen*, Math. Nachr., B. 49, N. 1-6, 1971, 173-185.

[82] K. Kuratowski, *Topology*, v. I, Academic press, Warszawa, 1966.

[83] K. Lancaster, *Nonparametric minimal surfaces in R^3 whose boundaries have a jump discontinuity*, Internat. J. Math. Math. Sci., v. 11, n. 4, 1988, 651-656.

[84] M. Lavrentieff, *Sur une classe de representations continue*, Math. sb., v. 42, n. 4, 1935, 407-424.

[85] M.A. Lavrentiev, *On continuity of univalence functions in closed domains*, Dokl. AN SSSR, v. 4, 1936, 207-210 (in Russian).

[86] M.A. Lavrentiev, B.V. Shabat, *Problems of hydrodinamics and their mathematical models*, Nauka, Moscow, 1973 (in Russian).

[87] J. Lawrynowicz, *On a class of quasiconformal mappings with invariant boundary points, II, Applications and generalizations*, Ann. polon. math., v. 21, n. 3, 1969.

[88] N.A. Lebedev, *Area principle in the theory of univalent functions*, Nauka, Moscow, 1975 (in Russian).

[89] H. Lebesgue, *Sur le probleme de Dirichlét*, Rend. circ. mat. Palermo, **24**, 1907, 371-402.

[90] O. Lehto and K.I. Virtanen, *Quasiconformal Mappings in the Plane*, Second Edition, Springer-Verlag, Berlin-Heidelberg-New York, 1973.

[91] J. Lelong-Ferrand, *Représentation conforme et transformations a intégrale de Dirichlet bornée*, Gauthier-Villars, Paris, 1955.

[92] B.E. Levickii, V.M. Miklyukov, *Reduced Modulus on a Surface*, Vestnik Tomsk. Gosud. Univ., n. 301, Tomsk, 2007, 87-91 (in Russian).

[93] B.E. Levickii, I.P. Mityuk, *'Narrow' theorems on space module*, Dokl. AN SSSR, v. 248, n. 4, 1979, 780-783 (in Russian).

[94] V.P. Luferenko, G.D. Suvorov, *On concept of a prime end body in the Caratheodory theory*, in 'Metr. voprosy theor. func. i otobr.', 'Metr. vopr. teor. func. i otobr.', n. III, Naukova dumka, Kiew, 1971, 71-79 (in Russian).

[95] Y. Li and L. Nirenberg, *The Distance Function to the Boundary, Finsler Geometry, and the Singular Set of Viscosity Solutions of Some Hamilton-Jacobi Equations*, Communications of Pure and Applied Mathematics, v. LVIII, 2005, 85-146.

[96] P. Li and J. Wang, *Finiteness of disjoint minimal graphs*, Math. Res. Letters, v. 8, 2001, 771-777.

[97] J. Maly, *Sufficient Conditions for Change of Variables in Integral*, in 'Proc. on Analysis and geometry', Izd-vo Inst. of math., Novosibirsk, 2000, 370-386.

[98] O. Martio, V. Miklyukov, *On existence and uniqueness of degenerate Beltrami equation*, Complex Variables, v. 49, 2004, 647-656.

[99] O. Martio, V.M. Miklyukov, and M. Vuorinen, *Critical points of A−solutions of quasilinear elliptic equations*, Houston Journal of Mathematics, v. 25, n. 3, 1999, 583-601.

[100] O. Martio, V.M. Miklyukov, and M. Vuorinen, *Relative distance and boundary properties of nonparametric surfaces with finite area*, Journal of Mathematical Analysis and Applications, v. 286, n. 2, 2003, 524-539.

[101] O. Martio, V. Ryazanov, U. Srebro, and E. Yakubov, *BMO-quasiconformal mappings and Q-homeomorphisms in space*, Prep. 288, Dept. Math. of Helsinki Univ., 2001, 24 pp.

[102] O. Martio, V. Ryazanov, U. Srebro, and E. Yakubov, *On the boundary behavior of Q-homeomorphisms*, Prep. 318, Dept. Math. of Helsinki Univ., 2002, 12 pp.

[103] V.G. Maz'ya, *On regularisity on the boundary of solutions of elliptic equations and conformal maps*, Dokl. AN SSSR, v. 152, n. 6, 1963, 1297-1300 (in Russian).

[104] E.J. McShane, *Parametrizations of saddle surfaces, with applications to the problem of Plateau*, Trans. Amer. Math. Soc., **35**, 1933, 716-733.

[105] D.E. Men'shov, *Les conditions de monogénéité*, Act. sci. et ind., 1936, v. 329, 1-52.

[106] V.M. Miklyukov, *On ε-quasiconformal mappings on a ball onto a ball*, in 'Metr. voprosy theor. func. i otobr.', n. I, Naukova Dumka, Kiew, 1969, 140-162 (in Russian).

[107] V.M. Miklyukov, *Some estimates of a conformal mapping of a domain to a strip*, Dokl. AN SSSR, v. 223, n. 3, 1975, 295-297 (in Russian).

[108] V.M. Miklyukov, *On some boundary problems of the conformal mapping theory*, Sib. math. j., v. 18, n. 5, 1977, 1111-1124 (in Russian).

[109] V.M. Miklyukov, *Two theorems on boundary properties of a minimal surface in the nonparametric form*, Math. zametki, v. 21, n. 4, 1977, 551-556 (in Russian).

[110] V.M. Miklyukov, *On the conformal type of surfaces, Liouville's theorem and Bernstein's theorem*, Sov. Math. Dokl., v. 19, n. 5, 1978, 1150-1153.

[111] V.M. Miklyukov, *On a new approach to Bernstein's theorem and related questions for equations of minimal surface type*, Math. USSR Sb., v. 36, n. 2, 1980, 251-271.

[112] V.M. Miklyukov, *On the asymptotic properties of subsolutions of quasi-linear equations of elliptic type and mappings with bounded distortion*, Math. USSR Sb., v. 39, 1981, 37-60.

[113] V.M. Miklyukov, *Some singularities in the behavior of solutions of equations of minimal-surface type in unbounded domains*, Math. USSR sb., v. 44, n. 1, 1983, 61-73.

[114] V.M. Miklyukov, *Some questions of the qualitative theory of minimal surface type equations*, in 'Boundary problems of the mathematical physics', Naukova Dumka, Kiew, 1983 (in Russian).

[115] V.M. Miklyukov, *Remarks on the lenght and area principle*, in the G.D. Suvorov book 'Generalized lenght and area principle in the theory of maps', Naukova Dumka, Kiew, 1985, 252-260 (in Russian).

[116] V.M. Miklyukov, *On critical points of solutions of maximal surface type equations in the Minkowski space*, Teoriya otobr. i pribl. func., Naukova Dumka, Kiew, 1989, 112-125 (in Russian).

[117] V.M. Miklyukov, *Stagnation zones of solutions to the Laplace-Beltrami equation in long strips*, Siberian Advances in Mathematics, v. 12. n. 4. 2002. 1-17.

[118] V.M. Miklyukov, *Isothermic coordinates on singular surfaces*, Sbornik: Mathematics, v. 195, n. 1, 2004, 65-84.

[119] V.M. Miklyukov, *Relative Lavrent'eff Distance and Prime Ends on Non-parametric Surfaces*, Ukrainian Mathematical Bulletin, v. 1, n. 3, 2004, 353-376.

[120] V.M. Miklyukov, *On some applications of the M.A. Lavrentiev relative distance*, Dokl. RAN, v. 402, n. 4, 2005, 448-451 (in Russian).

[121] V.M. Miklyukov, *On a property of the height function of multi dimensional Plateau problem*, Math. i prikladn. anal., n. 2, Izd-vo Tyumenskogo univ., Tyumen', 2005, 110-119 (in Russian).

[122] V.M. Miklyukov, *Area distortion under conformal maps of surfaces*, Trudy seminara po vectorn. i tenzorn. anal., n. XXVI, Izd-vo MGU, Moscow, 2005, 238-249 (in Russian).

[123] V.M. Miklyukov, *Speed of Approximation to Degenerate Quasiconformal Mappings and Stability Problems*, Complex Analysis and Potential Theory, Proceedings of the Conference Satellite to ICM 2006, World Scientific Publishing Co. Pte. Ltd., 2007, 33-45.

[124] V.M. Miklyukov, S.S. Polupanov, R.A. Tarapata, *On stabilization of gas dinamics equation solutions*, Dokl. RAN, v. 388, n. 5, 2003, 596-598 (in Russian).

[125] V.M. Miklyukov, G.D. Suvorov, *On existence and uniqueness of quasiconformal maps with unbounded characteristics*, in 'Issled. po sovrem. teorii funct. and its appl.', Naukova Dumka, Kiew, 1972, 45-53 (in Russian).

[126] J. Milnor, *On deciding whether a surface is parabolic or hyperbolic*, Amer. Math. Monthly, v. 84, n. 1, 1977, 43-46.

[127] Ch.W. Misner, K.S. Thorne, J.A. Wheeler, *Gravitation*, W.H.Freeman and Company, San Francisco, 1973.

[128] I.P. Mityuk, *Generalized reduced module and some of its applications*, Izv. vusov, Mathem., n. 2, 1964, 110-119 (in Russian).

[129] I.P. Mityuk, *On quasiconformal maps in the space*, Dopov. AN USSR, n. 8, 1964, 563-566 (in Ukrainian).

[130] V.N. Monachov, *Boundary problems with free boundaries for elliptic systems equations*, Nauka, Novosibirsk, 1977 (in Russian).

[131] M. Morse, *Topological methods in the theory of functions of a complex variables*, Princeton Univ. Press, Princeton, 1947.

[132] F. Morgan, *Geometric Measure Theory. A Beginners's Guide*, Second Edition, San Diego-Boston ets, Academic Press, 1995.

[133] C.B. Morrey, *On the solutions of quasi-linear elliptic partial differential equations*, Trans. Amer. Math. Soc., v. 43, 1938, 126-186.

[134] S. Müller, V. Sverák, *On surfaces of finite total curvature*, J. Differential Geometry, v. 42, 1995, 229-258.

[135] J. Nash, C^1-*isometric imbeddings*, Ann. Math., v. 60, 1954, 383-396.

[136] S.R. Nasyrov, *The metric space of Rimann surfaces over the sphere*, Russian Acad. Sci. Sb. Math., v. 82, n. 2, 1995, 337-356.

[137] J.C.C. Nitsche, *On new results in the theory of minimal surfaces*, Bull. Amer. Math. Soc., v. 71, n. 2, 1965, 195-270.

[138] O.A. Oleinik, G.A. Iosif'yan, *Sent-Venan principle analog and uniqueness of solutions of boundary problems in unbounded domains for parabolic equations*, Uspehi math. nauk, v. 31, n. 6, 1976, 142-166 (in Russian).

[139] O.A. Oleinik, G.A. Iosif'yan, *Boundary value problems for second order elliptic equations in unbounded domains and Saint-Venant's Principle*, Ann. Suola Norm. Sup. Pisa Cl. Sci. (4), n. 2, 1977, 269-290.

[140] O.A. Oleinik, E.V. Radkevich, *Analyticity and Liouville and Phragmén-Lindelöf type theorems for general elliptic system equations*, Math. sb., v. 95 (137), n. 1, 1974, 130-145 (in Russian).

[141] R. Osserman, *Minimal surfaces*, Uspehi Math. Nauk, v. 22, n. 4, 1967, 55-135 (in Russian).

[142] Y. Pan and T. Wolf, *A Remark on Unique Continuation*, The J. of Geometric Analysis, v. 8, n. 4, 1998, 599-604.

[143] V.L. Panteleev, *Theory of Earth Figure*, Izd-vo MGU, Moscow, 2000, 160 pp. (in Russian)

[144] V.I. Pelich, *Phragmén-Lindelöf theorems on minimal surfaces*, in 'Geometr. anal. i ego prilog.', Izd-vo Volgograd. univ., volgograd, 1999, 352-368 (in Russian).

[145] G. Piranian, *The distribution of prime ends*, Mich. Math. J., v. 7, 1960, 83-95.

[146] Ch. Pommerenke, *Boundary Behaviour of Conformal Maps*, Springer-Verlag, Berlin-Heidelberg-New York, 1992.

[147] I.I. Privalov, *Boundary properties of analytic functions*, GITTL, Moscov, 1950 (in Russian).

[148] Yu.G. Reshetnyak, *Stability theorems in geometry and analysis*, Nauka, Novosibirsk, 1982 (in Russian).

[149] Yu.G. Reshetnyak, *Two-dimensional manifolds of Bounded curvature*, in 'Geometry IV', Non-regular Riemannian Geometry, Springer-Verlag, Berlin-Heidelberg ets, 1993, 6-163.

[150] R.T. Rockafellar, *Convex analysis*, Princeton Univ. Press, Princeton, 1970.

[151] H. Rund, *The differential geometry of Finsler spaces*, Springer-Verlag, Berlin-Göttingen ets, 1959.

[152] C.A. Rogers, *Hausdorff measures*, Cambridge at the University Press, 1970.

[153] B. Rodin, S.E. Warschawski, *Extremal length and the boundary behavior of conformal mapping*, Ann. Acad. Sci. Fenn. Ser. AI Math, v. 2, 1976, 467-500.

[154] B. Rodin, S.E. Warschawski, *Extremal length and univalent functions II. Integral estimates of strip mappings*, J. Math. Soc. Japan, v. 31, 1979, 87-99.

[155] B. Rodin, S.E. Warschawski, *Estimates for conformal maps of strip domains without boundary regularity*, Proc. London Math. Soc., v. 39, 1979, 356-384.

[156] B. Rodin, S.E. Warschawski, *On the derivative of the Riemann mappings function near a boundary point and the Visser-Ostrowski problem*, Math. Ann., v. 248, 1980, 125-137.

[157] B. Rodin, S.E. Warschawski, *A necessary and sufficient condition for the asymptotic version of Ahlfors' distortion property*, Trans. Amer. Math. Soc., v. 276, 1983, 281-288.

[158] B. Rodin, S.E. Warschawski, *Conformal mapping and locally asymptotically conformal curves*, Proc. London Math. Soc., v. 49, 1984, 255-273.

[159] V.I. Ryazanov, *On compactness and metrizability of kernel spaces of manifolds*, Ukrainian Math. J., v. 50, n. 6, 1999, 944-951 (in Russian).

[160] V. Ryazanov, U. Srebro, E. Yakubov, *Degenerate Beltrami equation and radial Q-homeomorphisms*, Preprint 369, August 2003, Department of Mathematics, University of Helsinki.

[161] S. Saks, *Theory of the integral*, Warsaw, 1939.

[162] L. Sario, M. Nakai, C. Wang, L.O. Chung, *Classification Theory of Riemannian Manifolds*, Lectures Notes Math., 1977, 605 pp.

[163] M. Schiffer and D.C. Spencer, *Functionals of finite Riemann surfaces*, Princeton University Press, Princeton, New Jersey, 1954.

[164] L. Schwartz, *Analyse mathématique I*, Hermann, 1967.

[165] H.A. Schwarz, *Conforme Abbildung der Oberfläche eines Tetraeders auf die Oberfläche einer Kugel.* - J reine angew. Math., v. 70, 1869, 121-136.

[166] B.V. Shabat, *Introduction to the complex analysis*, Nauka, Moscow, 1969 (in Russian).

[167] E.V. Shikin, *On parabolicity of immersiability and hyperbolicity of non-immersiability twodimensional manifolds of the negative curvature*, Vestn. Moscow un-ta. Mathem. Mech., n. 5, 1990, 42-45 (in Russian).

[168] A.E. Shishkov, *Behaviour of generalized solutions of the Dirichlét problem for general quasilinear divergence elliptic equations at the boundary neighborhood*, Different. equat., v. 23, n. 2, 1987, 308-320 (in Russian).

[169] J. Serrin, M. Tang, *Uniqueness of Ground States for Quasilinear Elliptic Equation*, Indiana Univ. Math. J., v. 49, n. 3, 2000, 897-923.

[170] L. Simon, *Equation of mean curvature in 2 independent variables*, Pacific J. Math., v. 69, n. 1, 1977, 245-268.

[171] V.I. Smirnov, *Kurs vysshei matematiki*, v. 5, Fismatgiz, Moscow, 1959 (in Russian).

[172] S.L. Sobolev, *Some applications of the functional analysis in the mathematical physics*, Izd-vo Leningrad. univ., Leningrad, 1950 (in Russian).

[173] J. Spruck, *Two dimensional graphs over unbounded domains*, J. Inst. Math. Jussien, v. 1, 2002, 631-640.

[174] S. Stoilow, *Le cons sur les principes topologiques de la theorie de fonctions analytiques*, Cauthier - Villars, Paris, 1938.

[175] Yu.F. Strugov, *Quaiconformal in mean maps and extremal problems*, Parth 1, Omsk, 1994, 154 pp. - Dep. in VINITI 05.12.94, No 2786-94; Parth 2, Omsk, 1994, 114 pp. - Dep. in VINITI 05.12.94, No 2787-94 (in Russian).

[176] G.D. Suvorov, *Remarks to a M.A. Lavrentiev theorem*, Uch. zap. Tomsk. univ., v. 25, 1956, 3-8 (in Russian).

[177] G.D. Suvorov, *Families of plane topological maps*, Izd-vo SO AN SSSR, Novosibirsk, 1965 (in Russian).

[178] G.D. Suvorov, *Uniform stability of conformal maps of closed domans*, Ukrain. math. j., v. 20, n. 1, 1968, 78-84 (in Russian).

[179] G.D. Suvorov, *Generalized 'length and area' principle in the theory of maps*, Naukova Dumka, Kiew, 1985 (in Russian).

[180] G.D. Suvorov, *Prime ends and sequences of plane domains of plane maps*, Naukova Dumka, Kiew, 1986 (in Russian).

[181] A.V. Sychev, *Module and space quasiconformal maps*, Nauka, Novosibirsk, 1983 (in Russian).

[182] A.F. Tedeev, *Qualitative properties of solutions of the Neyman problem for a quasilinear parabolic equation of the height order*, Ukr. math. j., v. 45, n. 11, 1993, 1571-1579 (in Russian).

[183] O. Teichmüller, *Untersuchungen über konforme und quasikonforme Abbildung*, Dtsch. Math., n. 3, 1938, 621-678.

[184] V.G. Tkachev, A.N. Ushakov, *Fuglede theorem in Finsler spaces*, in 'The potential theory', Naukova Dumka, Kiew, 1991, p. 15 (in Russian).

[185] T. Toro, *Surfaces with generalized second fundamental form in L^2 are Lipschitz manifolds*, J. Differential Geometry, v. 39, 1994, 65-101.

[186] Yu.Yu. Trohimchyk, *Continuous maps and conditions of monogeneity*, GFML, Moscow, 1963 (in Russian).

[187] D.C. Ullrich, *Removable sets for harmonic functions*, Mich. Math. J., 1991, v. 38, n. 3, 467-473.

[188] N.X. Uy, *Removable set for Lipschitz harmonic functions*, Mich. Math. J., v. 37, 1990, 45-51.

[189] I.N. Vekua, *Generalized analytic functions*, GIFML, Moscow, 1959 (in Russian).

[190] G.M. Vergbickii, V.G. Maz'ya, *Asymptotic behaviour of second order elliptic equations close to boundary, I*, Sib. math. j., v. 12, n. 6, 1971, 1217-1249.

[191] G.M. Vergbickii, V.G. Maz'ya, *Asymptotic behaviour of second order elliptic equations close to boundary, II*, Sib. math. j., v. 13, n. 6, 1972, 1239-1271.

[192] L.I. Volkovyskii, *Quasiconformal mappings*, Izd-vo L'vov univ., L'vov, 1954 (in Russian).

[193] I.A. Volynec, *On distortion under mappings of the class BL*, Sib. math. j., v. 18, n. 6, 1977, 1259-1270 (in Russian).

[194] A.G. Vorob'ev, V.M. Miklyukov, *On some properties of subsolutions of minimal surface type equations* , Sib. math. j., v. 23, n. 1, 1982, 25-31.

[195] S.E. Warschawski, *Über das Randverhalten der Ableitung der Abbildungsfunction bei konformer Abbildung*, Math. Z., v. 35, 1932, 321-456.

[196] S.E. Warschawski, *On conformal mapping of infinite strips*, Trans. Amer. Math. Soc., v. 51, 1942, 280-335.

[197] S.E. Warschawski, *On the boundary behaviour of conformal mappings*, Nagoya Math. J., v. 30, 1967, 83-100.

[198] A. Weitsman, *On the growth of minimal graphs*, Indiana Univ. Math. J., v. 54(2), 2005, 617-625.

[199] H. Whitney, *Analytic extensions of differentiable functions defined in closed sets*, Trans. Amer. Math. Soc., v. 36, 1934, 63-89.

[200] T.H. Wolff, *Note on counterexamples in strong unique continuation problems*, Proc. of the Amer. Math. Soc., v. 114, n. 2, 1992, 351-356.

[201] T.H. Wolff, *A counterexample in a unique continuation problem*, Comm. in Analysis and Geometry, v. 2, n. 1, 1994, 79-102.

[202] V.A. Zhukov, *On a proof of the theorem on differential existence for W_p^1-functions*, Trudy Tomsk. un-ta, Ser. mech.-math., v. 189, 1966, 13-17 (in Russian).